Rethinking Corporate Sustainability in the Era
of Climate Crisis

Raz Godelnik

Rethinking Corporate Sustainability in the Era of Climate Crisis

A Strategic Design Approach

Raz Godelnik
The New School
New York, NY, USA

ISBN 978-3-030-77317-5   ISBN 978-3-030-77318-2   (eBook)
https://doi.org/10.1007/978-3-030-77318-2

Cover illustration: © John Rawsterne/patternhead.com

This Palgrave Macmillan imprint is published by the registered company Springer Nature
Switzerland AG
The registered company address is: Gewerbestrasse 11, 6330 Cham, Switzerland

To Peleg, Shira, and Omer

PREFACE

This book came to life because of my growing frustration with the slow progress companies have made on sustainability issues together with a growing sense of urgency around the climate crisis. While I was frustrated with companies' inadequate action, I also felt the climate crisis can be a game changer for sustainability in business because of the level of risk it poses and the growing pressure from young activists to act on it. Adding to it the increased attention to social justice issues and the impacts of COVID-19, it appears like the time is ripe for transformational changes in corporate sustainability. So I decided to sit down and write how this story of transformation should unfold. I wanted to make the case not just for why companies need to move away from what I call sustainability-as-usual (Chapters 1–4), but also what the desired change should look like and how we can make it happen (Chapters 5–8).

The lenses I use to address these questions are grounded in the field of strategic design. My approach to strategic design considers it not just as an effective way to address "wicked problems," but also as a bridge connecting context and content, radical thinking and practicality, humans and systems, and the present with the future. While you will not find many explicit mentions of strategic design in the book, it is embedded in every page, bringing attention to system structures, employing a human-centered point of view, challenging preconceptions, reframing problems, emphasizing the power of narratives and mental models, and seeking out clarity while embracing ambiguity.

As mentioned in the book, I like Dan Hill's notion that strategic design understands the criticality of the invisible dark matter to the creation of visible matter. Similarly, this book focuses on the conditions that are critical to shaping corporate sustainability, suggesting that changing the focal point from the matter (corporate sustainability practices) to the dark matter (e.g., the mental model that underlies corporate sustainability) may be a more effective way to transform corporate sustainability. This is not an easy task, but that is where strategic design comes into play, guiding us through the challenging journey of unlearning the obsolete mental models of yesterday and today, and giving us the confidence to step into the unknown.

My intention was to provide in this book a vision that is both radical and practical. This tension between having one's head in the clouds and keeping one's feet on the ground is not easy to grapple with and sometimes can be even painful. However, given the urgency of the climate crisis as well as the failure of companies so far to address it effectively, I believe that bringing these two opposing forces together is necessary to open up new pathways for change.

Finally, I would like to thank everyone who has supported me throughout this work, including The New School, Parsons School of Design, the School of Design Strategies (SDS), the wonderful research assistants who have helped me plant the seeds for this book over the last couple of years, particularly throughout the work on the Sandbox Zero project (www.sandboxzero.co): Melissa Brody, Paula Kawakami, Ruchi Thite, Silvia Garuti, and Dominique Pierre-Louis. I want to thank Dominique Pierre-Louis for her thoughtful comments on the manuscript and Paula Kawakami for her masterful work on the figures included in this book. Last but not least, I want to thank my family, whose love and support enabled me to complete this book.

New York, USA   Raz Godelnik

CONTENTS

ABOUT THE AUTHOR

Raz Godelnik is an Assistant Professor of Strategic Design and Management at Parsons School of Design—The New School in New York, where he explores sustainable business models and design solutions for the climate crisis. He currently serves as the Associate Director of the Strategic Design and Management BBA Program.

For over a decade, Raz has been working on corporate sustainability issues as entrepreneur, writer, researcher, and teacher, focusing on the connections between innovation, sustainability, business, and design strategies. In his latest project, Sandbox Zero (www.sandboxzero.co), Raz explores different strategies to move away from the current state of corporate sustainability, which he describes as "sustainability-as-usual." This project came out of Raz's frustration with dominating sustainability frameworks that do not seem adequate to address the environmental and social challenges we face, and particularly, the climate crisis.

Raz is the co-founder of two green startups (Hemper Jeans and Eco-Libris) and has written extensively on issues related to sustainable business, climate change, and strategic design. He holds an M.B.A. from Tel Aviv University and a B.A. in Communication and Economics from the Hebrew University.

LIST OF FIGURES

LIST OF TABLES

CHAPTER 1

Welcome to Business-As-Usual

Abstract This chapter introduces the theme of the first part of the book (Chapters 1–4), which is a critique of the sustainability efforts in companies during recent decades. These efforts, however, have not happened in a vacuum, but evolved out of a mode I describe as business-as-usual, which is the focus of this chapter. Its starting point is Milton Friedman's 1970 essay, in which he made the case that the only responsibility of companies is to maximize shareholder value. The chapter then moves to present the neoliberal economic thinking that Friedman and others harnessed to shape the narrative of a shareholder-centered era in business. It goes on to further explain the nature of business-as-usual by articulating the mental model dominating it: shareholder capitalism. Embraced by academia, business, courts, the media, and others, shareholder capitalism has become extremely influential, and even amid growing criticism of its merits, it continues to this day to play a critical role in shaping the business zeitgeist.

Keywords Business-as-usual · Milton Friedman · Shareholder capitalism · Sustainability · Shareholder primacy · Mental model

1

EVERYTHING HAS CHANGED
AND NOTHING HAS CHANGED

Business for Social Responsibility (BSR) is one of the leading organizations in the field of sustainable business, with a "network of more than 250 member companies and other partners" and a goal "to build a just and sustainable world."[1] In 2012, it published its 20th-anniversary report, which included the following point: "As we reflect on the past 20 years, it seems that everything has changed, and nothing has changed" (BSR 2012, 3). The report suggested that while immense progress has been made on sustainability in business, it is still insufficient. "By many objective measures, progress is far too slow" (3) explained the report.

Such critiques about the slow and/or insufficient progress on sustainability in business are not uncommon. Lacy et al. (2019), for example, suggested that while "over the last decade, sustainability has become firmly established on the corporate agenda...the new era of integrated sustainability, once confidently predicted by CEOs, has perhaps never seemed so far away." As C. B. Bhattacharya (2020), Chair in Sustainability and Ethics at the Katz Graduate School of Business at the University of Pittsburgh, wrote in his book *Small Actions, Big Difference*: "Sadly, my research led me to conclude that the current corporate sustainability movement is unsustainable". Furthermore, a 2020 report issued by the United Nations Global Compact pointed out that "Sustainable Development Goals (SDGs) provide a robust framework for addressing many of these inequities, but the evidence is clear that we are not on track to meet them. While there have been bright spots of progress in a number of areas, advancement toward the SDGs has been slow or even reversed" (Russell Reynolds Associates and UN Global Compact 2020, 5).

These critical voices suggest there may be a need to reevaluate the state of sustainability in business, especially in light of the rising sense of urgency on climate change. This book proposes the concept of "sustainability as usual" to describe the current state of sustainability in companies, which represents a shift from the state of "business as usual" but is still grounded in and defined by the latter. This chapter explores the concept of "business as usual," while the next one looks at "sustainability as usual" and the relationships between these two modes, which have evolved side by side in the last fifty years or so.

---

[1] See https://www.bsr.org/en/.

At its essence, the pursuit of sustainability in business so far has been about doing business differently: adopting a triple bottom line and considering the interests of multiple stakeholders instead of focusing only on profits and shareholder interests; giving priority to long-term rather than short-term considerations; and conducting business responsibly instead of doing whatever it takes to make money. The focus of these efforts has been very clear: moving away from business-as-usual thinking to a new mindset that offers a different vision for doing business. To understand this quest, one first needs to be familiar with what business-as-usual is all about. This is the starting point of this book's journey.

The Ideas that Have Shaped Business-As-Usual

The term "business as usual" conveys the normal course of things, or "an unchanging state of affairs despite difficulties or disturbances."[2] When it pertains to economic activity, it represents the prevalent economic thinking about firms and how they operate (Dyllick and Muff 2016). You can think of business-as-usual as the sum of all of our working assumptions about the ways in which companies conduct business, thus reflecting the dominant view about the essence of companies. At its core, business-as-usual mirrors a mindset[3] that is shaped by reality and is shaping it at the same time. To paraphrase economist Robert Frank (1988), "our beliefs about the nature of business help shape the nature of business itself. What we think about businesses and their possibilities determine what they will become."[4]

For the business world, it was Milton Friedman's 1970 article, "The Social Responsibility of Business Is to Increase Its Profits" that helped shape the understanding of what business-as-usual is all about, at least

---

[2] See https://www.lexico.com/en/definition/business_as_usual.

[3] In this book I adopt Heuer's (1999) approach and use the terms *mental model and mindset* more or less interchangeably to describe the lenses people use to process information. I concur with Heuer's notion that "the concept of mental model is likely to be better developed and articulated than a mindset" (4) and thus have some preference to using the term "mental model," which will be further explored later in this chapter.

[4] Frank's precise words were: "our beliefs about human nature help shape human nature itself. What we think about ourselves and our possibilities determines what we aspire to become" (Frank 1988, xi).

for the last fifty years. In less than 3000 words, Friedman articulated and championed the case for neoliberal economic thinking like no other. Published in *The New York Times Magazine* in September 1970, Friedman's piece made a clear case that companies should have only one consideration in mind—profit maximization.[5] He pointed out that a company is not responsible for its activities because it is an "artificial person," so the discussion should actually focus on the people who run the company. First and foremost, these managers are responsible to the owners of the company who, according to Friedman, are most likely to be interested in making as much money as possible. For Friedman, any deviance from this doctrine, or any talk about companies' "social conscience" was nothing but socialism, which he considered as the wrong path for business and society overall.

Friedman's case for shareholder primacy, whereby the sole purpose of the company is to maximize its shareholder wealth, was reinforced six years later with another influential piece, "Theory of the Firm," by Michael Jensen and William Meckling (1976). In a lengthy academic paper, the two economists suggested, just like Friedman, that managers are agents of the shareholders (principals) and thus are obligated to maximize shareholders' wealth only. Jensen and Meckling focused on the optimization of the agency relationships between shareholders and managers to ensure that the latter fulfill their contractual obligations to the former. At the core of their agency theory, as Bower and Paine (2017) pointed out, "is the assertion that shareholders own the corporation and, by virtue of their status as owners, have ultimate authority over its business and may legitimately demand that its activities be conducted in accordance with their wishes" (52).

The work of Jensen and Meckling helped to make the agency theory an important part of the business zeitgeist. Together with Friedman and other neoliberal economists, they helped shape the narrative of a shareholder-centered era, wherein all business activity is geared toward the maximization of shareholder value. While the ideas in these influential writings focused on how to conduct business effectively, they also

---

[5] Friedman made this point initially in his 1962 book *Capitalism and Freedom*, where he wrote that in a free economy "there is one and only one social responsibility of business—to use its resources and engage in activities designed to increase its profits so long as it stays within the rules of the game, which is to say, engages in open and free competition, without deception or fraud" (Friedman 1962, 133).

echoed a broader, neoclassical economic worldview that was grounded in a strong belief in free-market capitalism. Friedman explained the benefits of such a system in his *New York Times Magazine* article: "In an ideal free market resting on private property, no individual can coerce any other, all cooperation is voluntary, all parties to such cooperation benefit or they need not participate" (Friedman 1970, 132).

The neoliberal economic theory also had a clear and strong view on human nature, embracing the notion of the economic man (or *Homo Economicus*), who is driven by self-interest and makes decisions rationally based on utility maximization (Tittenbrun 2013). The idea of "economic man," which goes back to the work of John Stuart Mill in the nineteenth century, was adopted and further developed as an economic argument by the Chicago School (Vriend 1996). This influential neoclassical economic school of thought, which rose to fame with Chicago University economists like Friedman, George Stigler, and others, assumed that economic behavior reflects rational self-interest, suggesting that the choices people make are based on cold-hearted analysis of choices in search of the most advantageous one.

This approach was clearly reflected in explanations of people's motivations for working at firms. For example, Jensen (who did his Ph.D. at the Chicago School of Business) suggested that "it is inconceivable that purposeful action on the part of human beings can be viewed as anything other than responses to incentives" (1994, 41). This perspective may explain why Jensen and Meckling (1976) made a such strong case for offering managers equity-based incentives to ensure their interests are aligned with those of the shareholders.

Another important part of neoliberal free-market economic thinking that would manifest itself in the mode of "business-as-usual" is a limited power of the government. In his seminal book *Capitalism and Freedom*, Friedman (1962) laid out his idea of what government should be about at its core: "Its major function must be to protect our freedom both from the enemies outside our gates and from our fellow-citizens: to preserve law and order, to enforce private contracts, to foster competitive markets" (10–11).

It is interesting to consider that Friedman rejected the involvement of firms in "social" purposes, which he considered to be the government's job, but at the same time, he sought to reduce government intervention to minimum, making the case that it is rarely justified (Tilman 1976).

As to companies lobbying to make "the rules of the game" more favorable for them, Friedman was not supportive of it publicly (Cosans 2009), although there is evidence he offered the opposing view (i.e., support of corporate lobbying) in private meetings (Wilcke 2004). Overall, it seems like Friedman believed the answer to concerns about corporate lobbying lies in ensuring minimum intervention of the government, or in other words: "The less government could do to help or harm businesses, the less reason businesses would have to try to influence the government" (Fox 2018).

The aforementioned ideas of the neoliberal economic thinkers have played an important role in shaping the business landscape in the global North over the last decades.[6] The belief in shareholder primacy, profit maximization, rational actors, minimal government interference and the power of free markets helped frame the state of "business as usual" in the corporate world, becoming an integral part of business education and practice.

It is noteworthy that these ideas have become increasingly influential, even though they are not mandated by law, and sufficient evidence exists to question their efficacy. Take shareholder primacy for example. As corporate law scholar Lynn Stout explains in her book *The Shareholder Value Myth*: "U.S. corporate law does not, and never has, required public corporations to 'maximize shareholder value'" (Stout 2012, 23). Most corporate law scholars share this view, suggesting that shareholder primacy is a norm, not the law (Rhee 2018). More specifically, as Smith and Rönnegard (2016) claim, shareholder primacy established itself in practice as a social norm that business managers embrace because they believe it is their legal duty, have incentives that are often tied to share performance, or fear being penalized for not doing so.

Social norms are very powerful. Defined by Posner (1997) as "rule that is neither promulgated by an official source, such as a court or a legislature, nor enforced by the threat of legal sanctions, yet is regularly complied with" (365), they guide people's actions in many ways. Lessig (1998) offers a framework ("the new Chicago school theory"), in which social norms constitute one of four key constraints directing

---

[6] It should be noted that while this chapter refers for the most part to historical events and developments that took place in the U.S., these trends had shaped the world of business beyond the American business community due to the influence of American firms and neoliberal economic thinking on the global markets over the last decades.

people's behavior, along with law, markets (economics), and architecture (i.e., how things are designed). He suggests that each of these four forces has a regulating mechanism to constrain and guide behavior—markets use price, the law uses legal sanctions, and architecture uses built-in features. For social norms, Lessig points to the community enforcing its norms as the regulating mechanism.

Rhee (2018) uses Lessig's framework to make a broader case that shareholder primacy is a legal obligation,[7] not just a social norm, and that it is supported not only by the regulating mechanism of norms, but also by those of law, market, and architecture. "The workings of these three mechanisms, then, legitimizes and reinforces an existing norm of the business community" (1979), he writes. The point Rhee makes goes beyond the debate on the legal status of shareholder primacy. It is about showing the multi-level diffusion of the thinking on shareholder primacy, as well as how its prominent status in the discourse of corporate governance should be attributed to a number of forces working in tandem.

Lessig's framework, and Rhee's use of it, help us see the multiple elements that play a role in shaping the behavior of corporations and the relationships between these elements. However, to gain an even better understanding of what "business as usual" has been standing for in the last fifty years, we need to zoom out and consider mental models, not just social norms or legal rules. While social norms and mental models may seem similar, these two concepts are different, as mental models can provide a broader understanding, not just of human behavior, but also of systems behavior. As the World Bank explains in its 2015 World Development Report: "mental models, which need not be enforced by direct social pressure, often capture broad ideas about how the world works and one's place in it. In contrast, social norms tend to focus on particular behaviors and to be socially enforced" (World Bank 2015, 62).

---

[7] Rhee differentiates between legal duty and legal obligation, suggesting that shareholder primacy is "a legal obligation in the Hartian tradition: shareholder primacy is an important rule imbued with a "seriousness of social pressure," though it is not enforceable; it is recognized and institutionalized by courts" (Rhee 2018, 2006).

THE MENTAL MODEL OF BUSINESS-AS-USUAL

Mental models in general are representations of how something works, helping us understand and make sense of the world. Johnson-Laird, one of the leading scholars on mental models defines them as "internal representation[s] of a state of affairs in the external world" (1992, 932). Senge (1990) suggests that they are "deeply ingrained assumptions, generalizations, or even pictures or images that influence how we understand the world and how we take action" (8). "Very often, we are not consciously aware of our mental models or the effects they have on our behavior," (8) he adds. When many people share the same mental model, it helps them work together to "develop institutions, solve collective action problems, feel a sense of belonging and solidarity, or even understand one another" (World Bank 2015, 62).

From a system's point of view, mental models help us understand systems on a deeper level. Rouse and Morris (1986) suggest that they help people describe the purpose of the system (why it exists), explain its form and function, and predict potential future states. Another way to recognize the importance of mental models in systems is through the iceberg model. This popular systems thinking tool, which helps us see connections between events and the whole system, suggests that in systems, similarly to icebergs, we can see only a small part of the whole. The "tip of the iceberg" in systems represents events occurring around us that are grounded in a number of hidden levels: first patterns, then structures, and finally mental models. As Acaroglu (2020) points out: "moving from just seeing the events to understanding and challenging the mental models that reinforce the rest of the structure is one of the goals of a systems thinker" (105).

The iceberg model also suggests that the deeper we go the more our ability to change a system is enhanced, contrasting a very limited response on the "event" level ("react") with a much more meaningful one that could be achieved on the "mental model" level ("transform"). Scharmer (2009), who used the iceberg model in the development process of his "Theory U," explains the importance of addressing the mental model level to achieve systemic change: "If we want to upgrade our global economic operating system, we need to start by updating the thinking that underlies it; we need to update the essence of economic logic and thought" (11). Similarly, McKinsey's Keller and Schaninger (2019) report that according to their research, change programs of companies, which make the effort to recognize their "deep-seated mindsets," are far more successful compared to those of companies that fail to do so.

In the case of the business world, I consider "shareholder capitalism" to be the dominant mental model at the core of "business as usual." In other words, it is the lens through which the world of business is generally viewed and understood by executives, shareholders, and (almost) every other actor in this ecosystem, from business schools to media covering business to the courts. To use Kim's (1993) explanation of mental models, "shareholder capitalism" is for all of these actors "the context in which to view and interpret new material, and … determine how stored information is relevant to a given situation" (39).

This term *shareholder capitalism*, which is associated for the most part with the neoliberal thinking of Milton Friedman and the Chicago School about corporations and their role in society, could be interpreted in a number of ways. A narrower approach may view this term basically as another way of saying that the company's purpose is to maximize shareholder value (Bebchuk and Tallarita 2020; Gordon 1999). A broader approach may suggest this term denotes "a system driven by the interests of shareholder-backed and market-fixated companies" (McCann and Berry 2017, iii).

I concur with the latter (i.e., the broader) interpretation of shareholder capitalism. I see it as a "big tent" term, encompassing the centrality of shareholders in the firm, a belief that "business and markets are powerful institutions for efficient resource allocation" (Fürst 2017, 60), and the view that optimizing firms to maximize shareholder value is "good for economic vibrancy, both at the firm level and, in principle, for the economy overall" (Davis 2010, 321). The synthesis of these three elements creates a clear picture of the corporation as an efficient machine that makes everyone better off and contributes to the betterment of society—if and when it is left undisturbed to fulfill its obligations to its shareholders.

While one can debate whether shareholder capitalism is an ideology or just good economics, its role in shaping the business world in the last fifty years is less disputable. One of its most apparent manifestations has been the growing focus on companies' stock price as a key metric to measure their success. The prioritization of stock performance led not only to wrongly equating the value companies create with their value in the stock market (Mayer 2020), but also created pressure on companies to generate short-term returns, contributing to the financialization of the economy. As Foroohar (2016) points out: "Wealth creation within the financial markets has become an end in itself, rather than a means to the end of shared economic prosperity" (4).

Shareholder capitalism has become an intrinsic part of our socio-cultural fabric, which has an important role in shaping how people make sense of the world (Ringberg and Reihlen 2008). This function is evident in business, law, and economics schools, for example, where generations of business leaders, entrepreneurs, lawmakers, politicians, economists, and investors were educated over the last decades on the premise of shareholder capitalism. As Oxford business professor Colin Mayer (2018) writes: "Indeed virtually every MBA course begins from the premise that the purpose of business is to maximize shareholder value, and everything, and the rest of the course, follows from that. It reflects the power of ideas to influence behaviour to a point where many people now believe that the Friedman doctrine is a law of nature from which we are unable to escape" (2).

Academia was not the only institution to embrace shareholder capitalism with open arms. According to Hansmann and Kraakman (2001), the business and governmental elites also had a consensus around the "dominance of a shareholder-centered ideology of corporate law" (439). This narrative also dominated the media, where "shareholder-primacy rhetoric ... gave their readers a simple, easy-to-understand, sound-bite description of what corporations are and what they are supposed to do" (Stout 2012, 19). These trends echoed the rise of neoliberal economic views during this period of time, which considered corporations as the cornerstone of healthy market economies and championed their narrative of success: "In the West today the poor live better lives than all but the nobility enjoyed throughout the course of modern history before capitalism. Capitalism, plainly, has been the driving force behind this unparalleled economic and social progress" (Crook 2005, 6).

The growth in influence and the dominance of the shareholder capitalism mindset brought with it a spate of critiques, which offered multiple arguments against its merits. Bower and Paine (2017), for example, made the case that "the agency model's extreme version of shareholder centricity is flawed in its assumptions, confused as a matter of law, and damaging in practice" (52). In her book on the topic, Stout (2012) detailed the legal and economic flaws of the "shareholder primacy ideology," which she believes to be responsible for most of the problems corporations face. Mazzucato (2018) suggested that "maximizing shareholder value turned into a catalyst for a set of mutually reinforcing trends, which played up short-termism while downplaying the long-term view and a broader interpretation of whom the corporation should benefit" (166–67). Another harsh critique came from Mayer (2018), who wrote that "few social science theories are both so significant and misconceived as to threaten our existence but that is precisely what the Friedman doctrine is doing in the twenty-first century" (2).

The growing number of critics of shareholder capitalism showed not only its flaws, but also its perseverance. This mindset remained influential even after numerous corporate scandals, the 2008 economic crisis, and a growing attention to social and environmental issues that seemed to be only getting worse because of the focus on shareholder capitalism. While it appeared that shareholder capitalism was increasingly becoming anachronistic, it did not become obsolete, like many other symbols of its era. This longevity can be explained to some extent by the persistence of mental models in general. As the World Bank (2015) pointed out, "mental models, which may once have been well adapted to the situation at hand or may once have reflected the distribution of political power, can persist even when they are no longer adaptive or when the political forces that gave rise to them have changed" (63).

The persistence of shareholder capitalism did not mean that it was immune to challenges from alternative mental models trying to oust it from its dominant position. These efforts, which have focused on ways to make companies more responsible and sustainable, are aimed at moving business as far away from "business as usual" as possible. Yet as we will see in the following chapter, this process has proved to be extremely challenging in reality.

References

Acaroglu, Leyla. 2020. *Design Systems Change*. Disrupt Design LLC.

Bebchuk, Lucian A., and Roberto Tallarita. 2020. "The Illusory Promise of Stakeholder Governance." *Cornell Law Review* 106: 91–178. https://doi.org/10.2139/ssrn.3544978.

Bhattacharya, C. B. 2020. *Small Actions, Big Difference: Leveraging Corporate Sustainability to Drive Business and Societal Value*. New York, NY: Routledge.

Bower, Joseph L., and Lynn S. Paine. 2017. "The Error at the Heart of Corporate Leadership." *Harvard Business Review* 95 (3): 50–60.

BSR. 2012. "BSR at 20: Accelerating Progress." https://rb.gy/unew4u.

Cosans, Christopher. 2009. "Does Milton Friedman Support a Vigorous Business Ethics?" *Journal of Business Ethics* 87 (3): 391–99.

Crook, Clive. 2005. "The Good Company." *The Economist*, January 22.

Davis, Gerald F. 2010. "Is Shareholder Capitalism a Defunct Model for Financing Development?" *Review of Market Integration* 2 (2–3): 317–31.

Dyllick, Thomas, and Katrin Muff. 2016. "Clarifying the Meaning of Sustainable Business: Introducing a Typology From Business-as-Usual to True Business Sustainability." *Organization and Environment* 29 (2): 156–74.

Foroohar, Rana. 2016. *Makers and Takers: How Wall Street Destroyed Main Street*. New York, NY: Crown Business.

Fox, Justin. 2018. "The Purpose of the Corporation Isn't Lobbying." *Bloomberg*, June 12.

Frank, Robert H. 1988. *Passions Within Reason: The Strategic Role of Emotions*. New York: Norton.

Friedman, Milton. 1962. *Capitalism and Freedom*. Chicago: University of Chicago Press.

———. 1970. "The Social Responsibility of Business Is to Increase Its Profits." *The New York Times Magazine*, September 13.

Fürst, Michael. 2017. "Just When You Thought It Couldn't Get Worse, You Hear: 'The Business of Business Is Business'-Some Reflections on a Self-Fulfilling Prophecy and Alternative Perspectives on the Purpose of Companies." In *Creating Shared Value—Concepts, Experience, Criticism*, edited by Josef Wieland, 55–77. Springer.

Gordon, Jeffrey N. 1999. "Pathways to Corporate Convergence? Two Steps on the Road to Shareholder Capitalism in Germany: Deutsche Telekom and Daimler Chrysler." *Columbia Journal of European Law* 5 (219).

Hansmann, Henry, and Reinier H. Kraakman. 2001. "The End of History for Corporate Law." *Georgetown Law Journal* 89 (2): 439–68.

Heuer, Richards J. 1999. *Psychology of Intelligence Analysis*. Washington, DC: Center for the Study of Intelligence, Central Intelligence Agency.

Jensen, Michael C. 1994. "Self-Interest, Altruism, Incentives, and Agency Theory." *Journal of Applied Corporate Finance* 7 (2): 40–45.

Jensen, Michael C., and William H. Meckling. 1976. "Theory of the Firm: Managerial Behavior, Agency Costs and Ownership Structure." *Journal of Financial Economics* 3 (4): 305–60.

Johnson-Laird, P.N. 1992. "Mental Models." In *Encycolopeida of Artificial Intelligence*, edited by S. C. Shapiro, 2nd ed., 932–39. New York: Wiley.

Keller, Scott, and Bill Schaninger. 2019. *Beyond Performance 2.0: A Proven Approach to Leading Large-Scale Change*, 2nd ed. Hoboken, NJ: Wiley.

Kim, Daniel. 1993. "The Link Between Individual And Organizational Learning." *Sloan Management Review* 35 (1): 37–50.

Lacy, Peter, Pranshu Gupta, and Rob Hayward. 2019. "From Incrementalism to Transformation: Reflections on Corporate Sustainability from the UN Global Compact-Accenture CEO Study." In *Managing Sustainable Business: An Executive Education Case and Textbook*, edited by Gilbert G. Lenssen and N. Craig Smith, 505–18. Dordrecht: Springer Netherlands.

Lessig, Lawrence. 1998. "The New Chicago School." *The Journal of Legal Studies* 27 (S2): 661–91. https://doi.org/10.1086/468039.

Mayer, Colin. 2018. *Prosperity: Better Business Makes the Greater Good*. Oxford, UK: Oxford University Press.

———. 2020. "The Future of the Corporation and the Economics of Purpose." Finance Working Paper No. 710/2020. European Corporate Governance Institute. https://doi.org/10.2139/ssrn.3731539.

Mazzucato, Mariana. 2018. *The Value of Everything: Making and Taking in the Global Economy*. New York: PublicAffairs.

McCann, Duncan, and Christine Berry. 2017. "Shareholder Capitalism: A System in Crisis."

Posner, Richard A. 1997. "Social Norms and the Law: An Economic Approach." *American Economic Review* 87 (2): 365–69. https://doi.org/10.2307/2950947.

Rhee, Robert J. 2018. "A Legal Theory of Shareholder Primacy." *Minnesota Law Review* 122 (January). https://scholarship.law.umn.edu/mlr/122.

Ringberg, Torsten, and Markus Reihlen. 2008. "Towards a Socio-Cognitive Approach to Knowledge Transfer." *Journal of Management Studies* 45 (5): 912–35.

Rouse, William B., and Nancy M. Morris. 1986. "On Looking Into the Black Box. Prospects and Limits in the Search for Mental Models." *Psychological Bulletin* 100 (3): 349–63.

Russell Reynolds Associates, and UN Global Compact. 2020. "Leadership for the Decade of Action." https://unglobalcompact.org/library/5745.

Scharmer, Claus Otto. 2009. *Theory U : Leading from the Future as It Emerges : The Social Technology Presencing*. Berrett-Koehler.

Senge, Peter M. 1990. *The Fifth Discipline: The Art & Practice of the Learning Organization*. New York: Doubleday.

Smith, N Craig, and David Rönnegard. 2016. "Shareholder Primacy, Corporate Social Responsibility, and the Role of Business Schools." *Journal of Business Ethics* 134 (3): 463–78. https://doi.org/10.1007/s10551-014-2427-x.

Stout, Lynn. 2012. *The Shareholder Value Myth: How Putting Shareholders First Harms Investors, Corporations, and the Public*. San Francisco: Berrett-Koehler Publishers.

Tilman, Rick. 1976. "Ideology & Utopia in the Political Economy of Milton Friedman." *Polity* 8 (3): 422–42. https://doi.org/10.2307/3234360.

Tittenbrun, Jacek. 2013. "The Death of the Economic Man." *International Letters of Social and Humanistic Sciences* 11 (September): 10–34.

Vriend, Nicolaas J. 1996. "Rational Behavior and Economic Theory." *Journal of Economic Behavior & Organization* 29 (2): 263–85.

Wilcke, Richard W. 2004. "An Appropriate Ethical Model for Business and a Critique of Milton Friedman's Thesis:" *The Independent Review* 9 (2).

World Bank. 2015. "Thinking with Mental Models." In *World Development Report 2015: Mind, Society, and Behavior*. Washington, DC: World Bank. https://doi.org/10.1596/978-1-4648-0342-0.

CHAPTER 2

The Evolution of Sustainability-as-Usual

Abstract The chapter explores how sustainability has evolved in business over the last 50 years, reviewing developments in the understanding of the responsibilities and role of companies in society. During this period frameworks such as corporate social responsibility (CSR), stakeholder theory, and creating shared value (CSV) have been developed and implemented, reflecting the changes in companies' approach toward social and environmental issues. At the same time, while companies have made progress on these issues, they have continued to adhere to the idea that their primary role is to make money for their shareholders. As a result, this chapter suggests, corporate sustainability efforts fail to generate meaningful results. The effort to move companies away from an unsustainable "business-as-usual" mode, by promoting the premise of stakeholder capitalism as a framework that will replace shareholder capitalism, resulted in a new mode I call "sustainability-as-usual," which is dominated by a "shareholder capitalism 2.0" mental model. The chapter compares business-as-usual and sustainability-as-usual, acknowledging the shortcomings of the latter, including its inability to address climate change effectively.

Keywords Sustainability-as-usual · Corporate sustainability · Creating shared value (CSV) · Corporate social responsibility (CSR) · Triple bottom line · Stakeholder capitalism

R. Godelnik, *Rethinking Corporate Sustainability in the Era of Climate Crisis*, https://doi.org/10.1007/978-3-030-77318-2_2

The Rise of Corporate Responsibility

1970 was an important year for the journey presented in this book. Milton Friedman's *New York Times Magazine* article, which was discussed in length in the first chapter, was published in September that year in between two other key events: the first Earth Day took place in April and the U.S. Environmental Protection Agency (EPA) was created in December. These milestones represented two critical elements that will shape the context in which companies operate in the decades to come—activism and regulation.

On April 22, 1970, the first Earth Day brought millions of people across the U.S. together to raise awareness on various environmental issues, including air and water pollution and resource depletion. It was driven among other things by Rachel Carson's influential book *Silent Spring* (1962), as well as the 1969 burning of the heavily polluted Cuyahoga River in Ohio and a large oil spill in Santa Barbara, California, which brought more attention to the negative impacts of corporations on the environment (Latapí Agudelo et al. 2019).

These growing public concerns over the environment (as well as other political considerations) brought President Nixon to create the EPA later that year, thus forming one central governmental agency to provide better protection for public health and the environment. Furthermore, on the last day of 1970, Nixon signed the Clean Air Act (CAA of 1970), which is considered a landmark in the history of U.S. environmental law. Nixon saw it as the beginning of a new era, as we can discern from his remarks that day: "I think that 1970 will be known as the year of the beginning, in which we really began to move on the problems of clean air and clean water and open spaces for the future generations of America" (Nixon 1971).

1970 can also be considered "the year of the beginning" in terms of thinking about the roles and responsibilities of companies in society. This notion is true both in terms of the narrative "the business of business is business" and its counter-narrative, which suggests a much broader view on what "the business of business" should be all about. This debate clearly did not begin in 1970. Like Friedman's doctrine, which was developed and presented to the public earlier (Friedman 1962), the counter-narrative on corporate responsibility had a history of its own. For example, Carroll (2008) argued that while the history of corporate social responsibility (CSR) goes back to the Industrial Revolution, it is "mostly a product of the twentieth century, especially from the early 1950s" (19).[1] Still, 1970 can be considered a good starting point, given the importance of Friedman's article in shaping the debate over corporations, the new waves of environmental regulation and activism, and the growing efforts to articulate and codify an alternative mindset for business that took off in the 1970s.

In December 1970, the Committee for Economic Development (CED), an organization that included both business and education leaders, published a policy brief entitled *A New Rationale for Corporate Social Policy*. It included three papers making the case that "long-range corporate self-interest may be compatible with some types of social 'do-goodism'" (Baumol et al. 1970, vii). A year later, in another statement it issued, CED was more blunt about the expectations from companies: "Business functions by public consent, and its basic purpose is to serve constructively the needs of society-to the satisfaction of society" (Committee for Economic Development 1971, 11). It made the point that companies grew in size and influence in the twentieth century and that as they have grown, "they also have developed sizable constituencies of people whose interests and welfare are inexorably linked with the company and whose support is vital to its success" (19). The message was clear: shareholders are just one of a number of groups corporations are meant to serve.

Later on, in 1971, Keith Davis (one of the leading scholars in the field of corporate responsibility) gave a presentation at a conference at UCLA Los Angeles, in which he pointed out that social responsibility should go beyond just compliance with the law (see Davis 1973). Carroll (1979) developed this view further in a definition that provided additional clarity on the different facets of corporate social responsibility (CSR): "The social responsibility of business encompasses the economic, legal,

---

[1] This reference to the 1950s alludes in particular to Howard Bowen's (1953) seminal book *Social Responsibilities of the Businessman*, where he offered one of the first definitions of corporate social responsibility (or social responsibility as it was often referred to back then), suggesting it "refers to the obligations of businessmen to pursue those policies, to make those decisions, or to follow those lines of action which are desirable in terms of the objectives and values of our society" (Bowen 1953, 6). According to Carroll (1999), Bowen's novel work, which was the first one to seriously focus on the doctrine of social responsibility, makes him "the father of Corporate Social Responsibility" (270). More detailed accounts on the history of corporate social responsibility can be found in Carroll (1999, 2008) and Latapí Agudelo et al. (2019).

ethical, and discretionary expectations that society has of organizations at a given point in time" (500). According to Latapí Agudelo et al. (2019), this was "the first unified definition of corporate social responsibility" (6).

Carroll's definition evolved into a four-part framework for CSR, suggesting that "the CSR driven firm should strive to make a profit, obey the law, engage in ethical practices and be a good corporate citizen" (Carroll 2016, 6). In 1991, Carroll presented the four elements as a pyramid, making it known as "Carroll's CSR Pyramid," with economic responsibility at the base, followed by legal responsibilities, ethical responsibilities, and finally philanthropic responsibilities at the top (Carroll 1991). Carroll explained later on that "the pyramid should not be interpreted to mean that business is expected to fulfill its social responsibilities in some sequential, hierarchical, fashion, starting at the base. Rather, business is expected to fulfill all responsibilities simultaneously" (Carroll 2016, 6). From his point of view, corporate social responsibility was a whole made up of four equal elements that should be executed concurrently.

However, the picture was somewhat different as regards practice. First, the 1970s gave prominence to the legal responsibilities of corporations pertaining to the environment, especially in the U.S.—"a mountain of command-and-control regulation was passed during the decade of the 1970s, aimed at forcing companies to mitigate their negative impacts" (Hart 2010, 20). The growing regulatory pressures not only equated environmental responsibility with compliance, but also put companies in a defensive mode that changed for the most part in the 1980s, when they adopted a more cooperative approach (Hoffman 2001).

Clearly, the need to obey the law, or "play by the rules of the game" was central in the evolution of companies' understanding of their responsibilities in the 1970s, as the rules governing pollution, waste, hazardous materials, and so on in the U.S. became stricter. Yet possibly even more interesting was the corporate perception that viewed the changing regulation primarily through the lens of profit maximization. As Hart suggested, the growing command-and-control regulation prompted companies to develop what he called "the "Great Trade-Off Illusion"—the belief that firms must sacrifice financial performance to meet societal obligations" (Hart 2010, 21).

What we could already see in the 1970s was a clear hierarchy that remained in place in the decades to come, even as the concept of corporate social responsibility kept evolving: economic responsibilities (or considerations) always come first, then legal responsibilities, (which

companies try to keep as minimal as possible), and finally ethical responsibilities. Unlike Carroll's CSR Pyramid, in reality, philanthropic responsibilities were not truly part of the hierarchy: they seemed to be more of
a complementary component that mostly accompanied economic responsibilities. This three-level, hierarchical configuration had a formative role
in shaping the mental model that grew to challenge the shareholder capitalism mental model (i.e., shareholder capitalism 2.0, which is described
later on in this chapter) and in keeping these two mental models not too
far away from one another.

   While corporations adopted a more welcoming posture toward social
responsibility in the next decades, they kept adhering to the idea that
their key role was above all to make money. This view concurred with
Carroll's notion of economic responsibility as the most fundamental obligation business has to society: "At some point the idea of the profit
motive got transformed into a notion of maximum profits, and this has
been an enduring value ever since. All other business responsibilities are
predicated upon the economic responsibility of the firm, because without
it the others become moot considerations" (Carroll 1991, 41).

   This perspective was also clear to the people in charge of corporate
responsibility in companies. For example, Tim Mohin, who worked in
corporate responsibility in Intel, Apple, and AMD, wrote: "For-profit
enterprises survive or fail based on the value they deliver to the market.
As a CR worker, you need to understand your company's value proposition on a deep level and articulate how and why your CR programs add
to that value" (Mohin 2012, 232). In other words, corporate responsibility subjects itself to the financial considerations of the company, not the
other way around.

   This framing was evident in the evolving efforts of companies to
embrace voluntary measures to "green" their operations in the following
decades. Hoffman defines this stage as "enterprise integration... founded
on a model of business responding to market shifts to increase competitive positioning by integrating sustainability into preexisting business
considerations" (Hoffman 2018, 35). Hart further clarifies the economic
rationale of this shift, explaining that "by the late 1980s, it had become
clear that preventing pollution and other negative impacts was usually
a much cheaper and more effective approach than trying to clean up
the mess after it had already been made" (Hart 2010, 25). Overall,
companies' growing responsiveness to environmental and social issues
may also suggest an adoption of a more managerial and practical approach

to corporate responsibility, which Frederick (1978) describes as a shift from corporate social responsibility (CSR$_1$) to corporate responsiveness (CSR$_2$).

The 1980s and the 1990s also saw the rise of the stakeholder theory, which at its core asserts that a corporation should create value for all of its stakeholders, not just its shareholders. With Freeman's (1984) seminal book *Strategic Management: A Stakeholder Approach* and a large number of publications that followed it, this approach tried to provide a managerial-based alternative to shareholder capitalism. As Freeman and Velamuri (2006) explained: "An obvious play on the word 'stockholder', the approach sought to broaden the concept of business beyond its traditional economic roots, by defining stakeholders as 'any group or individual who is affected by or can affect the achievement of an organization's objectives'" (12).

The stakeholder theory sought to offer a more ethical approach, whereby one group of stakeholders (shareholders) is not preferred over the other groups. Its proponents suggested that it was also a better approach from a practical standpoint, given the growing understanding of the important role played by key stakeholder groups such as employees, suppliers, and others in creating value in organizations. A core element of the stakeholder theory is the pursuit of ways to overcome trade-offs and create mutual interests among different stakeholders, while aiming to "explore opportunities for installing positive links between stakeholder interests" (Hörisch et al. 2014, 331).

Freeman, who has been a leading figure in the development of the stakeholder theory, saw it as a new narrative for the world of business that would replace "the old story," which from his point of view was both deeply flawed in the first place and no longer a good fit for business in the modern age. Freeman saw the stakeholder theory at its core as an answer to the question of how to make business better. To do so, he claimed, you need to have a new story about the nature and purpose of business. In Freeman's mind, ethics and a sense of purpose are key to this new story, just as much as profit maximization was to the old one (Freeman 2017).

When Business Met Sustainability

Another trend that emerged in the 1980s and 1990s was sustainability in business, or corporate sustainability (CS). The term "sustainability" was popularized with the publication of the 1987 report "Our Common

Future" by the United Nation's World Commission on Environment and Development (WCED).[2] The WCED defined sustainable development in this report as "the development that meets the needs of the present without compromising the ability of future generations to meet their own needs" (WCED 1987, 43). The commission suggested that sustainable development requires taking environmental, social, and economic considerations into account concurrently (Montiel 2008), and companies started taking notice. Bansal (2005) framed these considerations as principles of environmental integrity, economic prosperity, and social equity, noting that "each of these principles represents a necessary, but not sufficient, condition; if any one of the principles is not supported, economic development will not be sustainable" (198).

One key element in the emergence of corporate sustainability was the notion of "beyond green"; that is, moving beyond earlier corporate efforts that had focused mainly on pollution prevention and product stewardship (i.e., considering the life cycle of a product when designing it) (Hart 2010). While such efforts demonstrated how environmentally responsible improvements in products and production processes can save companies money, these changes were incremental for the most part. Hart pointed out that sustainability offers a much greater upside potential for companies. The "sustainable enterprise," as he framed it, "represents the potential for a new private sector–based approach to development that creates profitable businesses that simultaneously raise the quality of life for the world's poor, respect cultural diversity, inspire employees, build communities, and conserve the ecological integrity of the planet for future generations" (Hart 2010, 17).

Others also saw the move beyond greening to sustainability as an important and strategic step for companies. Werbach (2009), for example, argued for the need to add a fourth dimension to sustainability—culture—which concerns the need to protect and value cultural diversity. He suggested that only a holistic strategy, which integrates the economic, social, environmental, and cultural dimensions of sustainability, can help companies thrive in an ever-changing world. Elkington also contributed significantly to promoting a holistic perception of sustainability with the term "triple bottom line" (TBL), which he coined in 1994. "In the

---

[2] The WCED was chaired by the former Norwegian Prime Minister Gro Harlem Brundtland and thus the report it produced is often referred to as the Brundtland report.

simplest terms," Elkington explained, "the TBL agenda focuses corporations not just on the economic value that they add, but also on the
environmental and social value that they add – or destroy" (Elkington
2004, 3).

Along those lines, a narrative has been constructed about the business
value of pursuing a sustainability agenda. According to Bansal (2005),
the growing commitment of corporations to sustainability was moti-
vated mainly by two main drivers—resource-based and institutional. The
latter refers to the social context of the firm's operations (e.g., social
norms), while the former relates to the resources used by the firm to
create value. This notion has evolved into further articulation of the business case for sustainability, whereby both academic and practice-based
experts show how sustainability efforts/practices are linked positively with
financial results (e.g., Eccles et al. 2014; Esty and Winston 2009; Fink
and Whelan 2016). An annual BSR/GlobeScan survey of the state of
sustainability in business suggested that key drivers of sustainability efforts
in business were changing over the years. For example, in 2009 cost
savings was considered to be a major driver, while in 2019 it was at the
bottom of the list. The top drivers in 2019 were reputational risk and
customer/consumer demand (BSR/GlobeScan 2009, 2019).

The pursuit of the business case for corporate sustainability has led
to the launch of reporting standards (e.g., GRI[3]) and frameworks (e.g.,
integrated reporting[4]), the formation of new sustainable business models
(Schaltegger et al. 2016), and a growing network that supports the
professionalization of this field (e.g., GreenBiz[5] and Sustainable Brands[6]).
It was also integrated into bolder visions, which attempted to reimagine
new formulations of value creation in business. In their seminal book
Natural Capitalism, for example, Hawken et al. (1999) described the
potential of having human and natural capitals fully valued, making
the case for a systemic design integration of environmental, social, and
economic considerations in business. About a decade later, Porter and
Kramer (2011) published their own seminal work on the concept of
creating shared value (CSV), which offered companies the opportunity

---

[3] See https://www.globalreporting.org/standards/.

[4] See https://integratedreporting.org/.

[5] See https://www.greenbiz.com/.

[6] See https://sustainablebrands.com/.

"to stop being trapped in an outdated approach to value creation" (64) by adopting a new model based on CSV. This principle, they wrote, is about "creating economic value in a way that also creates value for society by addressing its needs and challenges" (64). Porter and Kramer were offering a clear win–win for businesses, by connecting the dots between financial success and social progress.

Although Porter and Kramer suggested that shared value is not similar to sustainability or CSR (which they considered to be separate from profit maximization, unlike CSV), their concept became a prominent approach to making the business case for sustainability (de los Reyes and Scholz 2019). While it has been quickly endorsed by practitioners and scholars, CSV has also been criticized and contested for shortcomings such as ignoring tensions between economic and social targets, or naiveté about business compliance (Crane et al. 2014). De los Reyes and Scholz (2019) also expressed deep skepticism about the potential impact of CSV. "CSV can be expected, if successful, to do optimizing around the edges of those legacy businesses that are likely to continue contributing substantial harm to the environment," (787) they wrote. The researchers were critical of the business case for CSV specifically (and, by implication, the one for corporate sustainability), claiming it is based on optimization strategies (e.g., eco-efficiency) that may be beneficial for sustainability, but only on the margin. In other words, CSV may generate some progress, but not the transformative change we need to see in business; therefore, it is not truly the win–win model that CSV's proponents claim it to be.

The skepticism about the effectiveness of CSV may also shed light on broader issues pertaining to corporate sustainability. Consider for example the business case for sustainability. While the financial benefits of corporate sustainability have been demonstrated time and again in academic literature, in practice, it seems to be challenged, ignored, or overridden by other business considerations. Winston, who is one of the leading experts on this issue, pointed out in his book *The Big Pivot* (2014) that "to be realistic, large companies still view environmental and social issues in some less productive ways. Most top executives I engage with see sustainability as somewhat optional, incremental, short-term-focused (mostly about the 'gold' from eco-efficiency), and perhaps idealistic or naïve" (285). Bhattacharya (2020) added, based on his own research, "that most corporate sustainability endeavors fail because companies are either (a) unwilling to take action because they view sustainability as a costly add-on and a hindrance to short-term performance, or (b) unable to engage employees and integrate sustainability into their operations and stakeholders' lives" (14).

As these statements indicate, the problem with the business case for sustainability is not the lack of evidence or that the data are not clear enough, but possibly how sustainability is perceived by business leaders and executives. It may be directly connected to the aforementioned resource-based and institutional drivers of corporate sustainability (Bansal 2005). What we see could be the result of a lack of sufficient institutional pressure to support resource-based opportunities. In other words, while embedding sustainability in business offers financial advantages via its impact on both tangible and intangible assets (reducing supply chain risks, increasing customer loyalty and employee retention, etc.), it is not enough to compensate for the lack of meaningful support of the institutional environment.

Institutional environment, or the context in which companies operate, is a very powerful factor in the behavior of an organization. As Hoffman (2001) suggests in his work on the institutional history of corporate environmentalism, "organizational change is the product of institutional change" (9). While Hoffman asserts that institutional and resource-based (or technical) drivers for change may vary in their level of primacy, he also points out that a firm "is influenced by more than the technical influence of resource constraints ... it is also bound by social influences, embodied in rules, laws, industry standards, best established practices, conventional wisdom, market leadership and cognitive biases" (Hoffman 2001, 26).

The lack of sufficient institutional change to support a substantive shift in corporations does not mean necessarily that the ideas of win–win models as articulated even before Porter and Kramer by Prahalad (2006), Elkington (1994), and others are nothing but wishful thinking, but possibly that corporate sustainability needs reformulation, or reconsideration at a minimum.

From Sustainability to Sustainability-as-Usual

In June 2018, John Elkington, a key figure in the field of sustainable business, published an article in *Harvard Business Review*, wherein he "recalled" the term "Triple Bottom Line" (TBL), which he had coined almost 25 years earlier (Elkington 2018). Elkington made the case that while the concept became an important part of the sustainability

agenda/lexicon, it did not succeed in serving as a catalyst for system transformation, as he had hoped it would. Elkington saw a gap between his
vision for the TBL as an enabler of critical thinking about capitalism and
its actual use, as an approach to help corporations measure and manage
their environmental and social performance. At the end of the day, he
acknowledged, TBL evolved into an accounting tool that "has failed to
bury the paradigm of the single bottom line" (Elkington 2020, 32).

Elkington's recall was an important moment because it recognized
that the path that sustainability in business had been taking so far had
become unsustainable. While there had been progress on sustainability in
the business world, it was still too little, given the growing risk to people's
well-being and the health of the planet. As Elkington (2020) puts it:
"the sustainability sector's record in moving the needle on those goals
has been decidedly mixed. While there have been successes, our climate,
water resources, oceans, forests, soils and biodiversity are all increasingly
threatened" (30). To put it another way, Elkington asserts that decades of
work to create a new narrative of business, where companies are a force
for good in society and "part of solutions to societal problems rather than
a cause" (Freeman and Elms 2018, para. 17), have managed to generate
a mediocre outcome at best.

Building on Elkington's critique (and those of others), I suggest
that the effort to move the business world away from an unsustainable
"business as usual" mode ended up with a new mode (or state) I call
"sustainability as usual." In it, efforts to make businesses more sustainable
are the normal course of things, but simultaneously they are subjected to
the shareholder capitalism mental model (i.e., they can be pursued as long
as they are aligned in general with this mental model, or do not deviate
from it significantly). The notion of sustainability-as-usual echoes Bakan's
(2020) suggestion that while corporate sustainability efforts are real and
considerable, they have not changed companies fundamentally. "Making
money for themselves and their shareholders remains their top priority,
as it always has been. So while they might care about social and environmental values, they care only to the point such caring might cut into
profits" (27), he writes.

While its use was not common, the term "sustainability as usual" had
been employed in the past to describe different forms of insufficient
progress toward sustainability in business (e.g., Laszlo and Laszlo 2011;
Laszlo 2010). It had also been associated with the notion of weak sustainability (i.e., all forms of capital, including natural capital, are substitutable)

(Heikkurinen 2013). Elkington (2020) uses a similar term—"change as usual"—to characterize incrementalism in companies' change agendas. A 2013 publication by Volans, a think tank/consultancy Elkington co-founded, suggested that "many Change-as-Usual approaches were semi-revolutionary when they began—including corporate social responsibility, socially responsible investment and sustainability reporting. Where they succeed in being adopted by companies and industry sectors, however, they often become diluted by wider priorities" (Volans 2013, 16). Bhattacharya (2020) also offers a similar approach (although he does not use the term "sustainability as usual"), arguing that companies have failed to embrace sustainability and that the corporate mindset has not truly changed in the last fifty years.

Today we have growing evidence that while sustainability-as-usual does offer some progress on environmental and social issues, it is failing to move the needle. Examples of the failures of corporate sustainability-as-usual can be found with regard to progress made against the UN Sustainable Development Goals (SDGs) (Russell Reynolds Associates and UN Global Compact 2020), the response to the climate crisis (Coppola et al. 2019; Natural Capital Partners 2019; Dietz et al. 2020) and to human-rights violations (MSI Integrity 2020).

As the corporate response to the climate crisis is becoming more and more critical, it also emphasizes the limitations and risks of sustainability-as-usual. Deemed "the defining issue of our time" (United Nations 2019), climate change has brought up the need to act with greater urgency, as we already see growing evidence of its impacts unfolding (WMO 2019), including some more quickly than initially anticipated (e.g., Cheng et al. 2019; Leman 2019). In addition, the 2018 IPCC report provided a timeline with milestones that should be met to increase the chances of limiting global warming to 1.5 °C: "Global net human-caused emissions of carbon dioxide (CO_2) would need to fall by about 45 percent from 2010 levels by 2030, reaching 'net zero' around 2050" (IPCC 2018, 14). At the same time, even with all of the urgency around climate change, companies tend to move forward slowly, especially with climate reduction targets that are aligned with the Paris Agreement. One indication comes from *the 2021 State of Green Business report*, which suggested that current stated targets for global companies are "72 percent short of required emissions reductions to achieve the Paris Agreement"[7] (GreenBiz 2021, 90). Another indication of the slow progress is that, as of this writing, only about 1300 firms have adopted a science-based target.[8]

The reasons for this inadequate response vary, but in general, they can be associated with short-term cost-effectiveness considerations, the lack of legal requirements and insufficient pressure from key stakeholders. A UN Global Compact and Accenture Strategy report (2019), which shares insights from over 1000 global executives, suggests that economic constraints and business pressures play a key role in stalling corporate action on SDGs, including climate change. According to this study, 55% of the CEOs surveyed report that "pressure to operate with extreme cost-consciousness against investing in longer-term strategic objectives is a key trade-off that they are facing" (7). This perceived trade-off may also explain why while 86% of the CEOs surveyed acknowledged that action on climate is critical to achieving the SDGs, only 44% of the CEOs see that "a net-zero future is on the horizon for my company in the next ten years" (17).

Overall, sustainability-as-usual represents a theory of change whereby taking small steps (or making incremental progress) could eventually make a big difference. The problem with this theory is that incrementalism may work when the problem you face is static, but not when it keeps advancing, like climate change does. United Nations Secretary-General António Guterres (2018) reflected on this notion at the 2018 UN climate summit, suggesting that "climate change is running faster than we are and we must catch up sooner rather than later before it is too late" (para. 3). His words may suggest that even if companies feel they make some progress on this issue, they are actually moving backwards rather than forwards as the problem moves faster than the solution.

From my perspective, the key issue with sustainability-as-usual is the mental model dominating it, which was supposed to allow businesses to shift gears into a truly more sustainable and responsible mode. Some were hoping that stakeholder capitalism, which puts stakeholders at the center, would play the role of a transformational mental model.

---

[7] This analysis is based on data for companies included in the S&P Global 1200 index.
[8] See https://sciencebasedtargets.org/companies-taking-action/.

Freeman et al. (2007) described it as a model that is both pragmatic and libertarian in nature, focusing "on individuals voluntarily working together to create sustainable relationships in the pursuit of value creation" (311). Klaus Schwab, the founder of the World Economic Forum and a long-time proponent of stakeholder-based approach to business, claimed that stakeholder capitalism "offers the best opportunity to tackle today's environmental and social challenges" (Schwab 2019a, para. 2).[9] The World Economic Forum's Davos Manifesto 2020 offers a more detailed account of Schwab's vision, suggesting that "the purpose of a company is to engage all its stakeholders in shared and sustained value creation" (Schwab 2019c, para. 1).[10]

These propositions seem likely to remain wishful thinking for now. While stakeholder capitalism may have become a part of the business discourse, in practice, it looks like it has evolved to represent a "softer" version of the shareholder capitalism mental model. I call this mental model "shareholder capitalism 2.0": amplifying the narrative of stakeholder capitalism, while for the most part continuing to pursue shareholder capitalism.[11]

The Business Roundtable's 2019 Statement on the Purpose of a Corporation as an Example of Shareholder Capitalism 2.0

The Business Roundtable (BRT) is a leading corporate association, which includes CEOs of America's largest corporations. In August 2019, it released a "Statement on the Purpose of a Corporation" (Business

---

[9] Not everyone sees stakeholder capitalism this way, however. For example, Bebchuk and Tallarita (2020) make the case that stakeholderism offers an illusory promise that could actually hurt stakeholders by increasing managers' lack of accountability and by delaying or limiting valuable policy reforms.

[10] The original Davos Manifesto, which was published in 1973, opens with a very similar sentiment: "The purpose of professional management is to serve clients, shareholders, workers and employees, as well as societies, and to harmonize the different interests of the stakeholders" (Schwab 2019b, para. 2).

[11] Bebchuk and Tallarita's (2020) notion of "enlightened shareholder value" has some similarities to "shareholder capitalism 2.0." The researchers use the term to point out that stakeholder capitalism (or stakeholderism as they put it) is not that different from shareholder capitalism as it claims to be.

Roundtable 2019a), which was perceived as an endorsement of stakeholder capitalism (Huber et al. 2019; Vermaelen and Smith 2019). BRT has been publishing statements since 1978, and in every one released after 1997, it suggested that the purpose of business is to maximize shareholder value. The statement BRT released in 2019 had a presumably different message—this time it suggested that companies should commit to delivering value to all of their stakeholders, not just shareholders.[12]

This about-face seemed to represent more than just a change in language or a document update. Signed by more than 180 CEOs of large American corporations, the new statement echoed a desire of corporate America to display a clear shift in the narrative of business to one that is more stakeholder focused. It did not constitute an acknowledgment of the problems with shareholder capitalism as much as a reassurance of the positive role that companies play in society by creating more jobs, value, and prosperity for all. While for the most part it was hailed by commentators in the media as a positive and important shift (e.g., Ross Sorkin 2019), the statement also raised questions on the differences in practice between shareholder capitalism and the statement's version of stakeholder capitalism.

First and foremost, the voluntary nature of the commitments in the statement raises some questions about their true intent. For example, one of the commitments of BRT is "investing in employees. This starts with compensating them fairly and providing important benefits" (Business Roundtable 2019a, para. 5). At the same time, the statement does not ask for or demand laws to support this commitment, such as raising the federal minimum wage to a living wage, or providing all workers with paid parental leave. Summers (2019) argued that such regulation would actually make sense in terms of protecting against companies that may still be prioritizing shareholder interests. Giridharadas commented that "if the Business Roundtable is serious, it should tomorrow throw its weight behind legislative proposals that would put the teeth of the law into these boardroom platitudes" (Gelles and Yaffe-Bellany 2019, para. 30). The history of BRT suggests that this skepticism may be warranted, as the

---

[12] It should be noted that in 1981 BRT held a position somewhat similar to the one it adopted in 2019—according to Reich (2020), BRT adopted back then "a resolution noting that although shareholders should receive a good return, 'the legitimate concerns of other constituencies must have appropriate attention'" (102).

BRT has lobbied in the past against an increase in the minimum wage (Reich 2020).

Another indicator suggesting that the statement represents a mental model that offers very little change in comparison with shareholder capitalism is the lack of specific reference to climate change. One would expect that "a modern standard for corporate responsibility," as the statement is framed by BRT (Business Roundtable 2019b), would be very clear about one of the issues, if not *the* issue, defining now corporate responsibility, but there is not even a single mention of it in the statement. Instead, the statement states that "we respect the people in our communities and protect the environment by embracing sustainable practices across our businesses" (Business Roundtable 2019a, para. 12). This general and vague language does not seem to be accidental, as it allows BRT members to have their cake and eat it too when it comes to climate change.

Overall, it does not seem like the BRT statement deviates that much from the fundamentals of shareholder capitalism, where the corporation is expected to "to make as much money as possible while conforming to their basic rules of the society, both those embodied in law and those embodied in ethical custom" (Friedman 1970, 5). Similar to corporate social responsibility and sustainability frameworks, it does not offer to change the rules of the game: in essence, ethical responsibilities are to be maintained voluntarily. All it seems to be doing is recalibrating and expanding the notion of "ethical custom" to render it a better fit for this moment in history.

One BRT member who signed the statement is Larry Fink, the CEO of BlackRock, the world's largest financial asset manager. Every year Fink sends a public letter to the CEOs of all the companies in which Black-Rock invests on behalf of its clients.[13] His letters in the last couple of years have suggested very clearly that companies should adopt stakeholder capitalism. In his 2020 letter, for example, he wrote (in bold letters): "As I have written in past letters, a company cannot achieve long-term profits without embracing purpose and considering the needs of a broad range of stakeholders" (Fink 2020, para. 12). Unlike the BRT statement, Fink's latest letters also refer specifically to climate change, making it very clear that "there is no company whose business model won't be profoundly affected by the transition to a net zero economy" (Fink 2021, para. 14).

---

[13] The first letter, according to BlackRock's website (https://rb.gy/axl89i) was sent in 2012. Since then, with the exception of 2013, it is sent every year.

Coming from one of the most powerful people in the business world, Fink's letters have been widely accepted as a sign of a change toward a purpose-driven capitalism that is addressing climate change more responsibly (see e.g., Henderson and Serafeim 2020; Ross Sorkin 2020, 2021). At the same time, moving from words to action proved to be more difficult for Fink and BlackRock—for example, in May 2020 the company did not support a shareholder resolution asking it to prepare a board report on how its governance and management systems should be adjusted to fit the BRT statement[14] (As You Sow 2020). Similar to many other BRT members, it seems that Fink has been more comfortable overall with bold words than with bold action that deviates significantly from the status quo. Just like the other CEOs, he exemplifies a mindset that does not represent a significant withdrawal from "business as usual" as he would like us to think.

Sustainability-as-Usual vs. Business-as-Usual

Overall, BRT's statement provides a clear example of the state of sustainability-as-usual and the shareholder capitalism 2.0 mental model dominating it. In a sense, it is a culmination of the aforementioned CSR and other sustainability trends that evolved during the last decades in efforts to challenge business-as-usual and its dominant mindset (i.e., shareholder capitalism 1.0[15]). Figure 2.1 offers a framework to compare the mental models of sustainability-as-usual (i.e., shareholder capitalism 2.0) and business-as-usual (i.e., shareholder capitalism 1.0), which is inspired by Carroll's (1991) CSR Pyramid.

The framework suggests considerable similarities between the mental models in terms of structure, dynamics, and hierarchy. First, the mental models in both modes are grounded first and foremost in profit maximization, which dictates the nature of all other responsibilities. Second, while compliance with regulation is obligatory, companies work to limit any type of regulation to the minimum possible. Last but not least, in both mental models the need to act responsibly/sustainably remains optional.

[14] The shareholder resolution is available at https://www.asyousow.org/resolutions/2019/12/12/blackrock-corporate-purpose.

[15] To make a clearer sequence the mental model of "business as usual" is therefore reframed and presented hereinafter as "shareholder capitalism 1.0" instead of "shareholder capitalism."

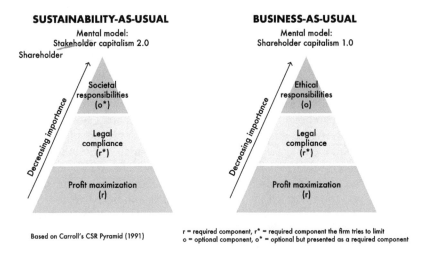

SUSTAINABILITY-AS-USUAL
Mental model:
Stakeholder capitalism 2.0

BUSINESS-AS-USUAL
Mental model:
Shareholder capitalism 1.0

Based on Carroll's CSR Pyramid (1991)

r = required component, r* = required component the firm tries to limit
o = optional component, o* = optional but presented as a required component

Fig. 2.1 Sustainability-as-usual vs. Business-as-usual

The key difference between the two mental models is in the third component—ethical responsibilities in business-as-usual vs. societal responsibilities in sustainability-as-usual. First, their scope is different—ethical responsibilities represent a much narrower scope, corresponding with Friedman's (1970) narrow view on the ethical duties of the firm, which should constrain profitmaking "only to the extent that ethical custom is reflected in the law" (McAleer 2003, 440).[16] On the other hand, societal responsibilities in the sustainability-as-usual mental model embody a broader understanding of the responsibilities of business toward society. It is in alignment with Bowen's broad approach to CSR (Acquier et al. 2011), which is illustrated, for example, in his definition of social responsibilities: "The obligations of businessmen to pursue those policies,

---

[16]While some scholars (e.g., Cosans 2009) have a broader interpretation of Friedman's notion of ethical duties of the firm, McAleer's (and others) narrow interpretation seems to be a better fit with Friedman's doctrine and its execution. In any event, the notion of narrow ethical duties in shareholder capitalism goes beyond Friedman's texts, reflecting a common critique among scholars about ethical responsibilities of corporations in the shareholder capitalism model (e.g., Bhattacharya 2020; Mayer 2018).

to make those decisions or to follow those lines of action which are desirable in terms of the objective and values of our society" (Bowen 1953, 6).

Second, while ethical responsibilities are relatively weak, with almost no impact on the two other components in the business-as-usual framework, societal responsibilities show greater strength as an element and have more impact on the two other elements of sustainability-as-usual, although not sufficient enough to change them. Examples include legal requirements to report on climate risks in the U.S.,[17] the creation of a benefit corporation as a legal structure,[18] the development of the shared value model, and the growing efforts to make a clearer and stronger business case for sustainability.

Finally, there is the issue of appearance and perception. Ethical responsibilities were clearly presented, and thus they were perceived as being a voluntary component of the business-as-usual mode. In other words, it was clear for companies that they could shape the ethical part of their behavior in whatever way they find suitable. While the application of this view has not changed in essence in the sustainability-as-usual mode, the way it is presented is very different. Endorsing stakeholder capitalism in principle, the sustainability-as-usual mode presents societal responsibilities as a required component of corporate behavior in the twenty-first century. Not only this assertion is unfounded, but it can also mislead the public about the efforts companies make to address sustainability issues and may even delay more effective measures (Bebchuk and Tallarita 2020).

Overall, BSR's notion of "everything has changed, and nothing has changed" (BSR 2012, 3) seems like a concise summary of "sustainability as usual"—it is an evolution from "business as usual" and we should not dismiss the changes that it has prompted in the business world. But at the same time, this progress should be evaluated not only vis-a-vis the starting point (i.e., business as usual), but also in relation to the changes that are needed in response to the climate crisis and even more importantly, the tight timeline for them to take place. The next two chapters discuss sustainability reporting (Chapter 3) and the circular economy (Chapter 4), two key parts of the work on sustainability in business, which

---

[17] See https://fas.org/sgp/crs/misc/IF11307.pdf.

[18] See list of states in the U.S. where benefit corporation legislation passed at https://benefitcorp.net/policymakers/state-by-state-status.

exemplify how sustainability-as-usual is failing to make sufficient progress. As these chapters demonstrate, the change sustainability-as-usual offers us may be nothing more than shuffling deck chairs on the Titanic.

REFERENCES

Acquier, Aurélien, Jean-Pascal Gond, and Jean Pasquero. 2011. "Rediscovering Howard R. Bowen's Legacy: The Unachieved Agenda and Continuing Relevance of Social Responsibilities of the Businessman." *Business and Society* 50 (4): 607–46.

As You Sow. 2020. "Fink Thinks BlackRock Already 'Operates in Accordance' with New Purpose of a Corporation—As You Sow." *As You Sow*, May 21. https://rb.gy/grudnh.

Bakan, Joel. 2020. *The New Corporation: How "Good" Corporations Are Bad for Democracy*. New York: Vintage Books.

Bansal, Pratima. 2005. "Evolving Sustainably: A Longitudinal Study of Corporate Sustainable Development." *Strategic Management Journal* 26 (3): 197–218.

Baumol, William J., Rensis Likert, Henry C. Wallich, and John J. McGowan. 1970. *A New Rationale for Corporate Social Policy*. New York: Committee for Economic Development.

Bebchuk, Lucian A., and Roberto Tallarita. 2020. "The Illusory Promise of Stakeholder Governance." *Cornell Law Review* 106: 91–178. https://doi.org/10.2139/ssrn.3544978.

Bhattacharya, C. B. 2020. *Small Actions, Big Difference: Leveraging Corporate Sustainability to Drive Business and Societal Value*. New York, NY: Routledge.

Bowen, Howard R. 1953. *Social Responsibilities of the Businessman*. New York, NY: Harper.

BSR. 2012. "BSR at 20: Accelerating Progress." https://rb.gy/unew4u.

BSR/GlobeScan. 2009. "State of Sustainable Business Poll 2009—Fact Sheet." https://rb.gy/tv1lup.

———. 2019. "The State of Sustainable Business in 2019." https://rb.gy/rmhv9p.

Business Roundtable. 2019a. "Statement on the Purpose of a Corporation." https://rb.gy/vbtxb4.

———. 2019b. "Business Roundtable Redefines the Purpose of a Corporation to Promote 'An Economy That Serves All Americans'," August 19.

Carroll, Archie B. 1979. "A Three-Dimensional Conceptual Model of Corporate Performance." *Academy of Management Review* 4 (4): 497–505.

———. 1991. "The Pyramid of Corporate Social Responsibility: Toward the Moral Management of Organizational Stakeholders." *Business Horizons* 34 (4): 39–48.

———. 1999. "Corporate Social Responsibility: Evolution of a Definitional Construct." *Business and Society* 38 (3): 268–95.

———. 2008. "A History of Corporate Social Responsibility." In *The Oxford Handbook of Corporate Social Responsibility*, edited by Andrew Crane, Matten Dirk, McWilliams Abagail, Moon Jeremy, and Donald S. Siegel. Oxford, UK: Oxford University Press.

———. 2016. "Carroll's Pyramid of CSR: Taking Another Look." *International Journal of Corporate Social Responsibility* 1 (1): 3. https://doi.org/10.1186/s40991-016-0004-6.

Carson, Rachel. 1962. *Silent Spring*. Boston/Cambridge, MA: Houghton Mifflin/Riverside Press.

Cheng, Lijing, John Abraham, Zeke Hausfather, and Kevin E. Trenberth. 2019. "How Fast Are the Oceans Warming?" *Science* 363 (6423): 128–29. https://doi.org/10.1126/science.aav7619.

Committee for Economic Development. 1971. "Social Responsibilities of Business Corporations." New York.

Coppola, Michela, Thomas Krick, and Julian Blohmke. 2019. "Feeling the Heat? Companies Are Under Pressure to Act on Climate Change and Need to Do More." *Deloitte Insights*, December.

Cosans, Christopher. 2009. "Does Milton Friedman Support a Vigorous Business Ethics?" *Journal of Business Ethics* 87 (3): 391–99.

Crane, Andrew, Guido Palazzo, Laura J Spence, and Dirk Matten. 2014. "Contesting the Value of 'Creating Shared Value'." *California Management Review* 56 (2): 130–53.

Davis, Keith. 1973. "The Case for and Against Business Assumption of Social Responsibilities." *Academy of Management Journal* 16 (2): 312–22.

Dietz, Simon, Rhoda Byrne, Dan Gardiner, Valentin Jahn, Michal Nachmany, Jolien Noels, and Rory Sullivan. 2020. "TPI State of Transition Report 2020." https://www.transitionpathwayinitiative.org/tpi/public ations/50.pdf.

Eccles, Robert G., Ioannis Ioannou, and George Serafeim. 2014. "The Impact of Corporate Sustainability on Organizational Processes and Performance." *Management Science* 60 (11): 2835–57. https://doi.org/10.1287/mnsc.2014.1984.

Elkington, John. 1994. "Towards the Sustainable Corporation: Win-Win-Win Business Strategies for Sustainable Development." *California Management Review* 36 (2): 90–100.

———. 2004. "Enter the Triple Bottom Line." In *The Triple Bottom Line: Does It All Add Up? Assessing the Sustainability of Business and CSR*, edited by Adrian Henriques and Julie Richardson. New York, NY: Earthscan.

————. 2018. "25 Years Ago I Coined the Phrase 'Triple Bottom Line.' Here's Why It's Time to Rethink It." *Harvard Business Review*, June. http://bit.ly/2NgNNYE.

————. 2020. *Green Swans: The Coming Boom in Regenerative Capitalism*. New York: Fast Company Press.

Esty, Daniel C., and Andrew Winston. 2009. *Green to Gold: How Smart Companies Use Environmental Strategy to Innovate, Create Value, and Build Competitive Advantage*. Hoboken, NJ: Wiley.

Fink, Carly, and Tensie Whelan. 2016. "The Sustainability Business Case for the 21st Century Corporation." New York, NY.

Fink, Larry. 2020. "Larry Fink's Letter to CEOs: A Fundamental Reshaping of Finance." *BlackRock*, January. https://rb.gy/vs0a3w.

————. 2021. "Larry Fink's 2021 Letter to CEOs." *BlackRock*. https://rb.gy/terps5.

Frederick, William C. 1978. "From CSR1 to CSR2: The Maturing of Business-and-Society Thought."

Freeman, R. Edward. 1984. *Strategic Management: A Stakeholder Approach*. Boston: Pitman Publishing Inc.

————. 2017. "The New Story of Business: Towards a More Responsible Capitalism." *Business and Society Review* 122 (3): 449–65.

Freeman, R. Edward, and Heather Elms. 2018. "The Social Responsibility of Business Is to Create Value for Stakeholders." *MIT Sloan Management Review*, January 4.

Freeman, R. Edward, Kirsten Martin, and Bidhan Parmar. 2007. "Stakeholder Capitalism." *Journal of Business Ethics* 74 (4): 303–14.

Freeman, R. Edward, and S. Ramakrishna Velamuri. 2006. "A New Approach to CSR: Company Stakeholder Responsibility." In *Corporate Social Responsibility: Reconciling Aspiration with Application*, edited by A. Kakabadse and M. Morsing, 9–23. Palgrave Macmillan, UK.

Friedman, Milton. 1962. *Capitalism and Freedom*. Chicago: University of Chicago Press.

————. 1970. "The Social Responsibility of Business Is to Increase Its Profits." *The New York Times Magazine*, September 13.

Gelles, David, and David Yaffe-Bellany. 2019. "Feeling Heat, C.E.O.s Pledge New Priorities." *The New York Times*, August 19. https://rb.gy/1pf5e6.

GreenBiz. 2021. "The State of Green Business 2021." https://rb.gy/htr4dz.

Guterres, António. 2018. "Secretary-General's Remarks at the Opening of the COP 24." *United Nations Secretary-General*. http://bit.ly/2Cy2zE1.

Hart, Stuart L. 2010. *Capitalism at the Crossroads*. 3rd ed. Upper Saddle River, NJ: Wharton School Publishing.

Hawken, Paul, Amory Lovins, and Hunter Lovins. 1999. Natural Capitalism: Creating the Next Industrial Revolution. *Journal of International Affairs*. Boston: Little, Brown and Co. https://doi.org/10.2307/24357805.

Heikkurinen, Pasi. 2013. "Reframing Strategic Corporate Responsibility: From Economic Instrumentalism and Stakeholder Thinking to Awareness and Sustainable Development." Aalto University.

Henderson, Rebecca, and George Serafeim. 2020. "Tackling Climate Change Requires Organizational Purpose." *AEA Papers and Proceedings* 110 (May): 177–80. https://doi.org/10.1257/pandp.20201067.

Hoffman, Andrew J. 2001. *From Heresy to Dogma: An Institutional History of Corporate Environmentalism*. Palo Alto, CA: Stanford University Press.

———. 2018. "The Next Phase of Business Sustainability." *Stanford Social Innovation Review*, no. Spring. https://ssir.org/articles/entry/the_next_phase_of_business_sustainability.

Hörisch, Jacob, R. Edward Freeman, and Stefan Schaltegger. 2014. "Applying Stakeholder Theory in Sustainability Management: Links, Similarities, Dissimilarities, and a Conceptual Framework." *Organization and Environment* 27 (4): 328–46.

Huber, Betty M., Joseph A. Hall, and Louis Goldberg. 2019. "Legal Implications of The Business Roundtable Statement on Corporate Purpose." *Harvard Law School Forum on Corporate Governance*, August 21.

IPCC. 2018. "Global Warming of 1.5 °C." https://www.ipcc.ch/sr15/.

Laszlo, Alexander, and Kathia Castro Laszlo. 2011. "Systemic Sustainability in OD Practice: Bottom Line and Top Line Reasoning." *OD Practitoner: Journal of the Organization Development Network* 43 (4): 10–16.

Laszlo, Kathia. 2010. "'Sustainability as Usual' Isn't Good Enough." *Triple Pundit*, May 17. https://www.saybrook.edu/blog/2011/03/23/sustainability-usual-isnt-good-enough/.

Latapí Agudelo, Mauricio Andrés, Lára Jóhannsdóttir, and Brynhildur Davídsdóttir. 2019. "A Literature Review of the History and Evolution of Corporate Social Responsibility." *International Journal of Corporate Social Responsibility* 4 (1): 1.

Leman, Jennifer. 2019. "Alarming Sonar Results Show Glaciers May Be Melting Faster Than We Expected." *Scientific American*. http://bit.ly/2Z1dVwd.

de los Reyes, Gastón, and Markus Scholz. 2019. "The Limits of the Business Case for Sustainability: Don't Count on 'Creating Shared Value' to Extinguish Corporate Destruction." *Journal of Cleaner Production* 221 (June): 785–94. https://doi.org/10.1016/j.jclepro.2019.02.187.

Mayer, Colin. 2018. *Prosperity: Better Business Makes the Greater Good*. Oxford, UK: Oxford University Press.

McAleer, Sean. 2003. "Friedman's Stockholder Theory of Corporate Moral Responsibility." *Teaching Business Ethics* 7 (4): 437–51.

Mohin, Timothy J. 2012. *Changing Business from the Inside Out: A Treehugger's Guide to the Corporate World*. San Francisco, CA: Berrett-Koehler Publishers.

Montiel, Ivan. 2008. "Corporate Social Responsibility and Corporate Sustainability: Separate Pasts, Common Futures." *Organization and Environment* 21 (3): 245–69.

MSI Integrity. 2020. "Not Fit-for-Purpose: The Grand Experiment of Multi-Stakeholder Initiatives in Corporate Accountability, Human Rights and Global Governance."

Natural Capital Partners. 2019. "Deeds Not Words: The Growth of Climate Action in the Corporate World." https://rb.gy/eblxat.

Nixon, Richard M. 1971. *Public Papers of the Presidents of the United States: Richard M. Nixon, 1970* . GPO. Washington, DC.

Porter, Michael E., and Mark R. Kramer. 2011. "Creating Shared Value." *Harvard Business Review* 89 (1/2): 62–77.

Prahalad, C. K. 2006. *The Fortune at the Bottom of the Pyramid: Eradicating Poverty Through Profits*. London: Dorling Kindersley.

Reich, Robert B. 2020. *The System: Who Rigged It, How We Fix It*. New York: Albert A. Knopf.

Ross Sorkin, Andrew. 2019. "How Shareholder Democracy Failed the People." *The New York Times*, August 20.

———. 2020. "BlackRock C.E.O. Larry Fink: Climate Crisis Will Reshape Finance." *The New York Times*, January 14. https://www.nytimes.com/2020/01/14/business/dealbook/larry-fink-blackrock-climate-change.html.

———. 2021. "BlackRock's Larry Fink Sets a Bolder Climate Goal in Annual Letter." *The New York Times*. https://rb.gy/u4alog.

Russell Reynolds Associates, and UN Global Compact. 2020. "Leadership for the Decade of Action." https://unglobalcompact.org/library/5745.

Schaltegger, Stefan, Erik G. Hansen, and Florian Lüdeke-Freund. 2016. "Business Models for Sustainability." *Organization and Environment* 29 (1): 3–10.

Schwab, Klaus. 2019a. "The Davos Manifesto: Towards a Better Kind of Capitalism." *World Economic Forum*, December 1.

———. 2019b. "Davos Manifesto 1973: A Code of Ethics for Business Leaders." *World Economic Forum*, December 2.

———. 2019c. "Davos Manifesto 2020: The Universal Purpose of a Company in the Fourth Industrial Revolution." *World Economic Forum*, December 2.

Summers, Lawrence H. 2019. "If Business Roundtable CEOs Are Serious About Reform, Here's What They Should Do." *The Washington Post*, September 2. https://rb.gy/vfnfbw.

UN Global Compact and Accenture. 2019. "The Decade to Deliver: A Call to Business Action." https://rb.gy/eb3mvk.

United Nations. 2019. "Climate Change." *United Nations*. https://www.un. org/en/sections/issues-depth/climate-change/.

Vermaelen, Theo, and N. Craig Smith. 2019. "BRT: A New View of Corporations and Capitalism." *INSEAD Knowledge*, August 23.

Volans. 2013. "Breakthrough: Business Leaders, Market Revolutions." London. https://volans.com/wp-content/uploads/2016/02/breakthrough-business-leaders-market-revolutions.pdf.

WCED. 1987. *Our Common Future*. Oxford: Oxford University Press.

Werbach, Adam. 2009. *Strategy for Sustainability: A Business Manifesto*. Boston, MA: Harvard Business Press.

Winston, Andrew. 2014. *The Big Pivot: Radically Practical Strategies for a Hotter, Scarcer, and More Open World*. Boston, MA: Harvard Business Review Press.

WMO. 2019. "WMO Statement on the State of the Global Climate in 2018." *World Meteorological Organization*. http://bit.ly/2N2xs9f.

CHAPTER 3

Sustainability Reporting: The Black Box

Abstract Reporting non-financial data has become a key element of corporate sustainability. Most public companies now offer regular updates on their progress toward voluntary sustainability commitments, reflecting a growing demand from investors and other stakeholders for greater transparency on sustainability-related issues. This chapter examines the evolution of sustainability reporting (SR), including the development of reporting standards and frameworks (e.g., GRI, SASB, IIRC), and their different approaches to SR. It questions SR's overall effectiveness in terms of advancing corporate sustainability meaningfully, looking under the hood of SR to study it from a systemic point of view. This inquiry considers three possible explanations for SR's shortcomings: applying a "mechanical" approach instead of a "biological" one, having little to no context, and adhering to a sustainability-as-usual mindset.

Keywords Sustainability reporting · ESG · Triple bottom line · Context-based sustainability · GRI · SASB

© The Author(s), under exclusive license to Springer Nature Switzerland AG 2021
R. Godelnik, *Rethinking Corporate Sustainability in the Era of Climate Crisis*, https://doi.org/10.1007/978-3-030-77318-2_3

Introduction

In 2018, John Elkington suggested the following observation in his introductory comments at the 5th International r3.0 (formerly Reporting 3.0) Conference: "it is almost as we are on a commercial airline and we're headed at full speed towards a very big mountain. Now, if you look at the dashboards, at the dials at the cockpit it is clear that we have enough fuel, we're going in the right speed, the temperature in the cabin is wonderful, the food is being served, the audio system is working quite nicely, but we're still headed for that mountain, and most of the reporting that we do doesn't take this mountain properly into account" (Elkington 2018).

Elkington's metaphor provides a good framing for sustainability reporting (SR),[1] which has become an important part of corporate sustainability. A 2014 EY report on SR suggested that "there are two key aspects of sustainability in business: reporting and strategy" (EY 2014, 4). Building on Elkington's metaphor, SR could be even considered the "black box" of corporate sustainability, as to some extent it is a recording of the latter's journey during the last few decades. Outlined in this chapter, the evolution of SR, its key challenges, and where it seems to be going will help shed more light on the current state of sustainability-as-usual, and on why companies need to move away from it.

Sustainability reporting can be defined as "the practice of measuring, disclosing and being accountable to internal and external stakeholders for organisational performance towards the goal of sustainable development...It involves reporting on how an organisation considers sustainability issues in its operations, and on its environmental, social and economic impacts" (European Court of Auditors 2019, 4). The Global Reporting Initiative (GRI) provides a more succinct definition: "A sustainability report is a report published by a company or organization about the economic, environmental and social impacts caused by its everyday activities" (GRI 2011).

The goals of SR, according to the GRI (2011), are to help companies understand, manage, and communicate their sustainability performance. Deloitte suggests that SR "can help an organisation to set goals, measure

---

[1] Other terms used interchangeably with SR are *corporate sustainability* (CSR) reporting and *environmental, social, and governance* (ESG) *reporting*.

performance, manage sustainability-related impacts and risks, and understand how it drives value for its stakeholders" (Deloitte 2020, 2). In general, SR can be used to support corporate sustainability practices in organizations and to communicate sustainability information to stakeholders interested in such data. Although SR has evolved to fulfill these functions better over the last three decades, it is still very much a product and reflection of the overall approach to sustainability in business, and as a result, it is bounded by the latter's overarching limitations. Thus, it is not too surprising that this chapter demonstrates how the mantra "it seems that everything has changed, and nothing has changed" (BSR 2012, 3) fits SR quite well.

THE EVOLUTION OF SR[2]

Disclosing non-financial information is not a novel idea. Companies have done it since the beginning of the twentieth century, focusing mainly on employee-related issues (Buhr et al. 2014). In the 1970s, employee reporting evolved into social responsibility disclosure, which was mostly produced by large multinational companies in Western Europe and the U.S. as a part of their annual reports, making this period "the decade of social reporting" (Buhr et al. 2014). This is not to say that environmental disclosure was completely neglected, as companies reported on a number of environmental and energy issues in their annual reports as well. For example, surveys by Ernst and Ernst of the annual reports of Fortune 500 companies between 1972 and 1978 found that the top three non-financial issues companies reported on were environmental issues, fair business practices, and community involvement (Ernst and Ernst 1978). This trend changed in the 1980s, which saw a decline in social reporting (largely attributed to the political shifts in the U.S. and the U.K. and the rise of free-market capitalism), followed by the emergence of environmental reporting at the end of the decade and in the early 1990s (Blowfield and Murray 2011).

The 1990s brought stand-alone environmental reports, a trend that grew slowly over the decade. A 1999 KPMG global survey found that 24% of the reporting companies published a separate environmental

---

[2] This part includes a brief history of SR only. For a more detailed account, see Mathews (1997) and Buhr et al. (2014).

report, compared with 17% in 1996 and 13% in 1993[3] (Kolk et al. 1999). According to the survey, the key drivers of this trend were the growing expectations of stakeholders for companies to act responsibly and transparently, along with the introduction of different mandatory environmental reporting requirements in a number of countries, including Denmark, the Netherlands, Norway, and Canada.

Another important development of the 1990s was the introduction of the triple bottom line (TBL). Coined and developed by John Elkington, the idea at its core was to shift the focus of individual companies, and eventually entire economies, from a single financial bottom line to "the simultaneous pursuit of economic prosperity, environmental quality, and social equity" (Elkington 1997, 397). The TBL was looking to provide clarity about what sustainable development actually means for business, offering an ambitious vision for the future: "No company pursuing sustainability – or operating in value chains where key actors have decided to move in this direction – will be able to ignore the triple bottom line. Companies will increasingly need to account for and report on their economic, environmental and social commitments, targets and performance" (SustainAbility and UNEP 1997, 3). To the credit of Elkington, TBL indeed managed to make a dent in the business world by becoming almost synonymous with corporate sustainability. Yet it also diluted the concept of sustainability, and may have led to higher levels of unsustainability (Milne and Gray 2013).[4]

The shortcomings of TBL, including "its measurement approach, its lack of integration across the three dimensions and its function as a compliance mechanism" (Sridhar and Jones 2013, 108), as well as its poor representation of what sustainability is about (Milne and Gray 2013), have been integrated into the fabric of SR, especially via the GRI. Formed in 1997, "with the goal of enhancing the quality, rigour, and utility of sustainability reporting" (Global Reporting Initiative 2002, i), the GRI has developed reporting guidelines and standards, offering companies a solid structure for sustainability reports that is grounded in the TBL. As the GRI explained (2002, 8), "this structure has been chosen because it reflects what is currently the most widely accepted approach to defining sustainability." With the continuous success of the GRI, which has grown

[3] KPMG conducted similar global surveys in 1993 and 1996.

[4] Elkington's own critique on the use of TBL can be found in Chapter 2.

in two decades to become the most popular SR framework,[5] TBL became not only part of the DNA of the GRI, but also of SR as a whole. GRI published the first version of its guidelines (G1) in June 2000 and continued to update them every few years, with the latest version (G4) being published in 2013. In 2016, the GRI transitioned from guidelines to standards by launching its sustainability reporting standards. While the GRI worked on improving its framework and spreading around the world, other frameworks were developed to complement and/or replace it, each one with its own raison-d'être, focus, and vision.

In 2000, the CDP (formerly the Carbon Disclosure Project) was launched. On behalf of many institutional investors, it asked companies to disclose environmental information, starting initially with their climate change impacts, and then adding water and forestry impacts. A decade later, the International Integrated Reporting Council (IIRC) was established, with the goal of promoting a holistic approach to reporting by integrating financial and non-financial information into one framework—the Integrated Report (IR). Three years later, the IIRC launched the <IR> Framework, with the ambitious goal of becoming "the next step in the evolution of corporate reporting" (IIRC 2013, 2).

Around the same time, another influential SR framework was created—the Sustainability Accounting Standards Board (SASB). Founded in 2011, its vision was to set clear reporting standards "to enable businesses around the world to identify, manage and communicate financially-material sustainability information to their investors" (SASB, n.d.). The SASB believes standards should be industry-based and should only include information that is financially material to investors. In 2018, the SASB issued reporting standards for 77 industries, identifying sustainability topics, which may be material for companies in each industry. These standards are meant to support both voluntary sustainability reports and mandatory disclosures.

Another key framework is TCFD, the Task Force for Climate-Related Financial Disclosures. Founded in 2015 by the Financial Stability Board (FSB), it was tasked to "develop voluntary, consistent climate-related financial disclosures that would be useful in understanding material risks related to climate change" (TCFD 2020, ii). In 2017, the TCFD released

---

[5] For example, 67% of N100 reports and 73% of G250 reports use the GRI framework (KPMG IMPACT 2020).

recommendations, which focused on four areas: governance, strategy, risk management, and metrics and targets. The TCFD focuses on assessing climate change-related opportunities and risks (including physical and transition risks) by asking companies to conduct scenario analysis to assess the long-term effects that climate change will have on them. While implementation of the TCFD's recommendations could be challenging,[6] they are gaining in popularity with companies (KPMG IMPACT 2020), investors (see, e.g., Fink 2020), and governments (see, e.g., Diaz-Rainey 2020).

The final piece in the SR's "alphabet soup" is SDGs, i.e., the U.N. Sustainable Development Goals. Adopted in 2015 by the U.N., the 17 goals address critical challenges, from climate change to extreme poverty and hunger, and are meant to be achieved by 2030. While a growing number of companies integrate SDGs into their sustainability reporting (KPMG IMPACT 2020), most of them struggle to figure out how to use them effectively. As PwC's Scott and McGill (2018) reported, although a large number of companies engage with and pledge commitments to the SDGs, "there remains a gap between companies' good intentions and their ability to embed the SDGs into actual business strategy" (6).[7] Furthermore, the fact that they were not developed as a SR framework for companies seems to make the SDGs less appealing for investors, who apparently show more interest in business-oriented SR frameworks such as TCFD and SASB (see, e.g., Fink 2020, 2021).

As the use of SR continued to grow, one issue that remained a source of debate has been the question of whether SR should be mandatory or voluntary; i.e., are regulators or market forces the ones to drive SR? According to *Carrots and Sticks*, a report published every few years on trends in SR regulation and policy, 54% of the SR provisions worldwide in 2020 were mandatory and 46% were voluntary. Interestingly, while this ratio was very similar to the one found in 2006, it also reflected a decline in the percentage of mandatory provisions when compared to 2010 and 2013 (72 and 65%, respectively). The 2020 report showed that

---

[6] Based on the findings of a TCFD survey asking companies about the level of difficulty implementing the four recommendations (TCFD 2020).

[7] An example to the difficulty to use SDGs can be found in a global survey PwC conducted, which found that only 1% of the surveyed companies "measured their performance against SDG targets" (Scott and McGill 2019, 7).

governments and stock exchanges were prominent forces in issuing SR requirements, with Europe leading the charge, followed by Asia. Overall, there is some evidence that mandatory SR may have more impact than voluntary SR (Christensen et al. 2019)—although, as Christensen et al. point out, "much will depend on the specificity, proper implementation and enforcement of the standards (including assurance)" (55).

The Battle of the Standards

In September 2020, five leading SR organizations—CDP, CDSB, GRI, IIRC, and SASB—published a joint document detailing their plan to work together, along with the TCFD, to create a unified SR framework. These organizations explained that they wanted to provide clear guidance on "how our frameworks and standards can be applied in a complementary and additive way" (CDP, CDSB, GRI, IIRC, and SASB 2020, 3). Their goal is to create a coherent global standard framework that could meet the needs of companies and the capital markets.[8] The idea of sustainable reporting standardization also receives greater attention from governmental and accounting bodies, such as the IFRS Foundation[9] and EFRAG[10] (Eccles 2020).

The shared statement was looking to end an ongoing schism lasting a decade or so between the different organizations that were working to establish the dominance of their point of view on the shape and form of SR (especially GRI, SASB, and IIRC). Accompanied by attempts to find common ground and ways to work together (e.g., the Corporate Reporting Dialogue[11] and the Impact Management Project[12]), these battles reflected different approaches among the SR organizations on issues such as the target audience of the report, the definition of materiality, and their general theory of change.

---

[8] Two months later the IIRC and SASB announced that they will merge into one organization in 2021 (Cohn 2020).

[9] International Financial Reporting Standards Foundation—see at https://www.ifrs.org/.

[10] European Financial Reporting Advisory Group—https://www.efrag.org/.

[11] See https://corporatereportingdialogue.com.

[12] See https://impactmanagementproject.com.

First, these organizations have different target audiences in mind, as the GRI sees all stakeholders as the target audience and the SASB and the IIRC have been focusing primarily on investors (Hess 2014). The differences in target audiences were also echoed in the organizations' definitions of materiality. This fundamental concept in accounting is widely used in financial reporting, where it offers guidance on the relevance of information for the report on the basis of its significance to the users, and it also became a key concept in SR. Similar to financial reporting, materiality in SR aims to ensure the clarity and usefulness of the report by setting processes in place to help decide what information should be disclosed within it (Unerman and Zappettini 2014). This notion was interpreted differently according to the organizations' overall approach to SR and their understanding of who their target audience is. We can see this aspect in particular by looking at the GRI and the SASB as examples.

Evolved from the SASB's core objective to provide investors with material sustainability information, its standards focus on financially material issues, which are those "that are reasonably likely to impact the financial condition or operating performance of a company and therefore are most important to investors" (SASB 2018). On the other hand, the GRI defines a material topic (i.e., one that needs to be included in the report) as one that "reflects a reporting organization's significant economic, environmental and social impacts; or that substantively influences the assessments and decisions of stakeholders" (GRI 2018, 13).

The multi-stakeholder approach of the GRI and the investor-centered one of the SASB also reflected very different theories of change. The GRI's theory suggests that its standards will help companies be more transparent about their environmental, social, and economic impacts: "this greater transparency will give companies and their stakeholders the information they need to make decisions for change. This changed behavior, toward reducing negative impacts and maximizing positive impacts, is an important part of realizing a sustainable global economy" (GRI 2016, 6). The idea is not just to provide stakeholders with needed sustainability information to inform their decisions, but also to ensure they are involved in the process and that the standards are developed with their input in the first place. According to the GRI, this inclusive approach will drive sustainable change in business.

In contrast, the SASB's theory of change is about igniting sustainable change at scale by aligning companies' and investors' interests. Once investors are provided with material information on a company's ESG performance, they can factor it into their decision-making/capital allocation processes, which in return incentivize companies to improve their ESG performance. This theory essentially suggests that investors are the most impactful stakeholder group in terms of incentivizing companies' sustainability performance; therefore, the goal of the standards is to create a positive feedback loop between them and the company.

The aforementioned shared statement published in September 2020, which Eccles (2020) described as "the most significant step in setting global ESG reporting standards taken to date" (para. 5), tried to overcome these differences and create a common language that all SR frameworks could adopt. One example was the use of the concept of "dynamic materiality" (see Kuh et al. 2020), which is a "nested" approach to materiality that could bridge the differences between those of the GRI and the SASB. These efforts to find a common ground came out of a sense of urgency that was shared by all five SR organizations. In the statement, they suggested that this is a pivotal moment, requiring "progress towards a more comprehensive solution for corporate reporting; one that is urgently needed to improve enterprises' contribution to sustainable development, to help address climate change and to enable more resilient, efficient financial markets" (CDP, CDSB, GRI, IIRC, and SASB 2020, 3). Yet what is still unclear is if SR can actually make all of this happen.

The Effectiveness of Sustainability Reporting

In its different formats, SR has provided much hope that it could advance corporate sustainability in a meaningful way. The assumption of SR proponents has always been that SR could help companies measure and improve their environmental and social impact (see, e.g., EY and Boston College Center for Corporate Citizenship 2013). Tang and Demeritt (2018) articulated this argument as follows: "behavioural change will occur because reporting will encourage firms to develop a deeper understanding of an issue or topic, while being forced to disclose their performance publicly will incentivize them to take steps to improve that performance" (453).

Examining this assumption will become more critical as SR gains in popularity. The many companies publishing SR reports both in the U.S. (90% of the S&P 500 companies in 2019; G&A Institute 2020) and worldwide (96% of G250 and 80% of N100 companies in 2020; KPMG

IMPACT 2020)[13] indicate a growing appetite for sustainability data, as well as the "normalization" of SR in the business world. Within only three decades or so, SR has become "widely recognized by financial stakeholders as a critical component of corporate reporting" (KPMG IMPACT 2020, 2). At the same time, we are still left with the question of the extent to which these reports help companies improve their sustainability performance.[14]

Maas and Vermeulen (2015) studied the theory of change behind SR (both mandatory and voluntary), in which a growing demand for SR (input) will lead eventually to improved sustainability performance (impact). Their findings suggested that the evidence supporting this theory of change is scant, while pointing out the general difficulty in establishing causality relationships between SR and non-financial performance, given the many factors influencing the latter. Robert Gray, one of the leading scholars in this field and a critic of SR, also reached a similar conclusion for non-mandatory SR, suggesting that "there is nothing in the literature to help us state categorically that voluntary disclosure reliably signals or influences social and environmental performance" (Gray 2006, 78). On the other end, Bednárová et al. (2019) found a positive correlation between the level of environmental data disclosure and environmental performance.[15]

With the growing focus on climate change and companies' emissions, a number of researchers have looked specifically at the impact of climate disclosure on companies' greenhouse gas (GHG) emissions. Some researchers found no impact of GHG emissions disclosure on companies' actual emissions (i.e., no material improvement of the latter because of the former; e.g., Matisoff 2013; Belkhir et al. 2017), while others showed different results (i.e., reduction in emissions due to climate disclosure;

---

[13] Another indication of the wide use of SR globally is provided by Corporate Register, a global online directory of SR. According to the website (https://www.corporatereg ister.com/), by September 2020, more than 20,000 companies published about 125,000 sustainability reports.

[14] Other notable impacts of SR that have been studied by researchers include the impact on companies' performance or market value. However, this chapter looks primarily at SR's impact on corporate sustainability performance, which is the focus of the book.

[15] The positive correlation was between an environmental reporting index, which was based on measuring the number of environmental KPIs disclosed in sustainability reports using GRI standards, and environmental score as retrieved from the Newsweek Green Rankings in 2017.

e.g., Downar et al. 2020). Then again, according to a recent extensive review of Zhang and Liu (2020) on the effects of carbon information disclosure (CID), most studies show that such disclosures have no or limited effect on emission reductions and ecological betterment.

A comprehensive global study published by The Conference Board in 2020 reiterated Zhang and Liu's conclusions. Looking into nearly 6000 companies, the report suggested that while SR is on the rise, "more disclosure does not necessarily translate into changing practices" (The Conference Board 2020a, 3). One example in the study for this gap between disclosure and practice is carbon emissions, where more companies reported on climate risks yet also emitted more CO_2. The lead author of the report suggested that this gap "is likely due, in part, to companies not yet integrating sustainability as an integral part of their business strategy, and thus they lack an incentive to change their practices" (The Conference Board 2020b, para. 4).

The abovementioned evidence suggesting that the rise of SR is not accompanied by considerable changes in companies' sustainability performance is troubling as it calls into question the notion that the mainstreaming of SR is an indication of actual progress in corporate sustainability and a real change in the ways in which companies create and deliver value (see, e.g., Robinson et al. 2019). Looking at the surface level may suggest that the failure of SR to support significant changes in corporate sustainability is due to issues with the reports, from lack of standardization to inadequate materiality. However, a deeper level of inquiry could offer a different point of view, whereby the main problem is not with the reports themselves, but with the system they are part of, which may set them up to fail in the first place.

Looking Under the Hood, or, Why SR Is Not Working

Tim Mohin, the then CEO of the GRI, stated the following in his testimony before the U.S. Congress: "Our theory of change boils down to the axiom that 'you manage what you measure.' All organizations run on data. By identifying, measuring—and most importantly reporting—about the most material ESG topics, these issues will be managed, and performance will improve" (Mohin 2019, 2). The notion that "you manage what you can measure" seems to be quite popular in SR circles, as it provides not only a good framework for change, but also a plausible explanation for any setbacks along the way. A lack of success in improving sustainability performance could always be tied to issues with reporting or measuring, which means that to receive better results an organization needs to have a better reporting system.

This approach could make sense to some degree, especially given the general dissatisfaction with the current disclosure system. For example, a McKinsey study found that "investors say they cannot readily use companies' sustainability disclosures to inform investment decisions and advice accurately" (Bernow et al. 2019, para. 1). According to the study, the main shortcomings that investors identified in current SR practices were incomparability, inconsistency, and lack of alignment in the standards. Investors want reports that are financially material, consistent, and reliable (i.e., audited). In addition, there are also concerns that reporting companies have embraced what Serafeim (2020) calls "a 'box-ticking' culture" when approaching ESG issues, instead of addressing them more strategically. Lastly, many companies also have complaints about the current SR system, according to a new proposed SR framework by the World Economic Forum (WEF), which aims to respond to these concerns (WEF 2020).

However, looking under the hood of SR may provide us with a somewhat different understanding of its failure to drive meaningful change in corporate sustainability. Taking a deeper look provides three possible explanations for SR's ineffectiveness: applying a "mechanical" approach instead of a "biological" one, having little to no context, and adhering to a sustainability-as-usual mindset.

Taking a Mechanical Rather Than a Biological Approach to Fixing SR

The current thinking on SR assumes conventional linear cause-and-effect relationships between measuring/disclosing and managing/improving sustainability impacts in organizations, thus applying more of a "mechanical" approach to corporate sustainability. This reductionist approach considers SR as a machine part that can be studied separately and then fixed when necessary, so that the machine (in this case corporate sustainability and/or the company) can work better.

This approach may not prove valuable in today's complex and unpredictable business environment. Reeves and Levin (2017) suggest that this environment requires us to think about companies in terms of complex adaptive systems (CASs) and to apply a "biological" approach, "which acknowledges the uncertainty and complexity of business problems and so addresses them indirectly" (para. 4).

According to John Holland (2006), CASs "have a large numbers of components, often called agents, that interact and adapt or learn" (1). The Health Foundation's extensive study of their use suggests that "in its most simple form, complex adaptive systems is a way of thinking about and analysing things by recognizing complexity, patterns and interrelationships rather than focusing on cause and effect" (The Evidence Centre 2010, 6). Clayton and Radcliffe (1996) emphasize "interaction with their environment and change in response to environmental change" (23) as a distinctive feature of CASs, and Espinosa and Porter (2011) add that a CAS approach reflects the notion of organisms co-evolving with their environment.

Applying a biological approach rather than a mechanical one implies not just addressing companies as CASs, but also corporate sustainability as a CAS in which SR is one element (or component). This proposition is consistent with a growing body of research suggesting that companies should apply a systemic approach in general, and a CAS approach in particular, to corporate sustainability (Porter and Derry 2012), sustainable business models (Dentoni et al. 2020), and sustainable innovation (Iñigo and Albareda 2016). It also goes hand in hand with the complex and adaptive character of both organizational systems (Smith and Lewis 2011) and sustainability (Gaziulusoy and Brezet 2015; Espinosa and Porter 2011), thus reflecting the notion that "a complexity approach better reflects today's multileveled dynamic of organizational life and offers fresh insights to sustainability-related dilemmas" (Porter and Derry 2012, 36).

Figure 3.1 shows each system as a part of a bigger system: corporate sustainability is part of a corporation that is part of a business environment. This hierarchical structure of nested systems (i.e., systems that are embedded in broader systems) is a key element in CASs (Reeves and Harnoss 2017). Gaziulusoy and Brezet (2015) underline the adaptive significance of such structures, due to the interaction formed between the different levels.

This difference between the mechanical and biological approaches is critical. A mechanical approach would focus primarily on repairing the part rather than assessing the whole for systemic shortcomings. As a result, it is more likely to promote a direct intervention to fix it, such

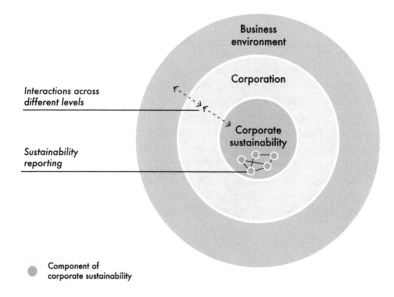

Fig. 3.1 Viewing sustainability reporting through the lenses of complex adaptive systems

as those offered by the five SR organizations (CDP, CDSB, GRI, IIRC, and SASB 2020) and the WEF (2020). On the other hand, a biological approach would take a more holistic view and may offer an indirect intervention that could help create an external change to address the problem (Faeste et al. 2019). Given the complexity of sustainability issues and climate change in particular, an indirect intervention could be far more effective than a direct one, as BCG's Faeste et al. (2019) noted: "To address a complex task...direct interventions (such as mandating individual behaviors) are unlikely to bring about the required change" (para. 25). At the same time, they write, "Indirect interventions...often prove to be more effective because they touch the deeper, more persistent drivers of behavior" (para. 25).

To conclude this point, the notion that SR could be fixed in isolation, without considering the CASs in which it is embedded, simply ignores system properties such as emergence, co-evolution, connectivity, and nestedness, and it is likely to fail. As Gaziulusoy points out: "sustainability can only be achieved using non-reductionist, dynamic systems thinking"

(Gaziulusoy 2011, para. 8). Although offering direct interventions such as refined materiality guidelines or better standardization may improve the quality of the reporting, they will probably not add to its effectiveness, given the lack of consideration of the bigger picture.

The Missing Context

Sustainability reports have become a celebration of continuous progress toward sustainability goals—for example, "we have also made steady progress toward using water more efficiently and to treat all wastewater in our production processes" (The Coca Cola Company 2020, 7). But is this touted progress meaningful? Is it progress at all? It is almost impossible to tell, when in most cases companies only make comparisons against their own goals. If one thing is missing from these reports, it is context.

Allen White, the GRI's co-founder, points out that "sustainability requires contextualization within thresholds. That's what sustainability is all about" (Baue 2014, para. 7). McElroy and van Engelen (2012) concur, suggesting that "sustainability measurement, management and reporting must be context based in order to be meaningful" (8). They explain that context in this case refers to "the background state of vital social and environmental resources in the world and what an organization's impacts on them ought to be (or not to be) in order to be sustainable" (2). They argue that the sustainability context is missing from both corporate sustainability and SR. Evidence for the lack of context in SR was provided in an extensive study that looked at 40,000 sustainability reports published by 12,000 companies between 2000 and 2014 (Bjørn et al. 2017). The researchers found that just 5% of them made a reference to ecological limits on any given year, and only 31 in total used these ecological limits to define different targets.

Given the growing understanding of the importance of staying within planetary and social boundaries to ensure we live in a safe and just space (see, e.g., Raworth 2017), there is a case to be made about the little value of a reporting system offering decontextualized measurements that ignore these boundaries. As Utting (2020) noted in a policy brief from the United Nations Research Institute for Social Development (UNRISD): "Even if improvements in both qualitative and quantitative dimensions of corporate performance are occurring, there is no way of knowing, under current reporting formats, whether such improvements are significant from the perspective of sustainable development" (2).

A context-based approach to sustainability was developed by McElroy (2008), first in his Ph.D. dissertation and later in a book co-written with van Engelen (2012). Context-based sustainability (CBS) considers the carrying capacities of multiple capitals (not just financial ones), as well as their organizational allocation. McElroy (2020) points out that CBS is based on "the measurement of impacts on multiple capitals relative to organization-specific norms, standards, or thresholds (NSTs) for what such impacts would have to be in order to be sustainable (empirically so)" (2). Baue adds that "enacting CBS enables organizations to practically apply the notion of thresholds and allocations to track their performance" (Baue 2019, 4).

The shortcomings of the current state of SR can be exemplified via McElroy's CBS metric: the sustainability quotient ($S = A/N$). This equation suggests that sustainability for an organization equals the actual impacts on the carrying capacity of vital capitals ("A") divided by its normative impacts ("N"). Currently, companies are mostly reporting on the numerator ("A") only, which has little to no value without the denominator. Moreover, this partial representation may provide an illusion of progress that could be more dangerous than no progress (Baue and Thurn 2020).

The slow but steady growth in the use of science-based targets (SBTs),[16] which Baue considers to be "the first major initiative in the corporate sphere to apply a thresholds-and-allocations approach" (Baue 2019, 21), shows CBS is not far-fetched after all. At the same time, SBTs are still the exception rather than the rule for sustainability disclosure. While the discourse on CBS and the implications for SR continue to evolve (see, e.g., Thurm 2017), in practice, decontextualized SR still dominates the space.

It's the Mental Model, Stupid

It may be easy to point a finger at the decontextualized sustainability reports or at the mechanical approach toward corporate sustainability as reasons for the ineffectiveness of SR. Yet if we look under the hood carefully, we may find that the state of SR is a reflection of the state of

---

[16] See at https://sciencebasedtargets.org/companies-taking-action/.

sustainability-as-usual and of the mental model dominating it. This viewpoint echoes the notion that "systems thinking requires that we recognize that in human-designed systems, repeated events or patterns derive from systemic structures, which, in turn, derive from mental models" (Monat and Gannon 2015, 19). In other words, understanding the ineffectiveness of SR requires a consideration of the broader environment in which it takes place, including the mental model manifested in the practice of SR. To do so, we will use the iceberg model, a popular systems thinking tool, which can help connect the dots between the publication of sustainable reports and the environment in which they are created (see Fig. 3.2).

Let us go through the different levels presented in the iceberg model:

Events: This level represents what we can actually see, or what is above the surface, to use the iceberg metaphor. In this case, the event is the failure of sustainable reports to support a meaningful improvement in companies' sustainability impacts in general, and climate impacts in particular. Attempts to fix the problem on this level could focus on enhancing the standardization, consistency, and comparability of the metrics used in the reports (see, e.g., CDP, CDSB, GRI, IIRC, and SASB 2020;

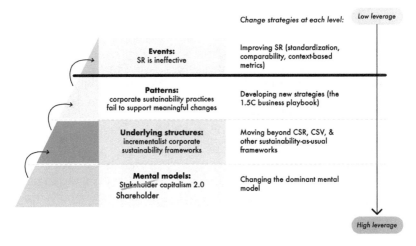

Fig. 3.2 Explaining sustainability reporting through the iceberg model

WEF 2020). A more advanced strategy on this level would look at using context-based metrics.[17]

Patterns: Looking below the event level, we try to identify patterns and trends over time, of which the abovementioned event is a part. Ultimately, sustainability reporting is not that different from the many other sustainability practices that companies fail to execute effectively. Therefore, we can consider SR's failure as part of a broader failure of companies to address sustainability adequately. As Bhattacharya (2020) notes, "when it comes to practicing and not just preaching sustainability – when companies have to develop and implement a sustainable business model for themselves – many struggle and most founder" (5). The response on this level could focus on adopting a new implementation strategy, such as the 1.5C Business Playbook (Falk 2020), or Bhattacharya's (2020) sustainability ownership model.

Underlying structures: On this level, we look for structures or foundations that cause the patterns we observe, which can be either physical or intangible (Kim 1999). This level "reveals how the different components are interconnected and affect one another" (Al-Homery et al. 2019, 953). The patterns of sustainability practice failures in business seem to be driven by the conceptual frameworks employed by companies to address sustainability challenges, such as corporate social responsibility (CSR) and creating shared value (CSV). These frameworks appear to be ineffective as remedies for shareholder capitalism (de los Reyes and Scholz 2019; Schneider 2019), as they result in practices that fail to produce meaningful outcomes and do relatively little to reduce the negative impacts of business activities. The response on this level would be to design a new framework for corporate sustainability.[18]

Mental models: Mental models[19] can be viewed as "systemic structure generators" because they provide the "blueprints" for these structures" (Kim 1999, 5). Systemic structures such as CSR and CSV are not created and disseminated in a vacuum. Glauner (2019), for example, suggests that they are driven by a more traditional economic mental model, which is built on the notion of the scarcity of resources and the competition

---

[17] See, for example, the context-based metrics offered by the Center for Sustainable Organizations, https://www.sustainableorganizations.org/context-based-metrics-public-domain/.

[18] An example of such a framework is presented in Chapter 6.

[19] An introduction to mental models can be found in Chapter 1.

over them. I refer to the mental model shaping the underlying structures as "shareholder capitalism 2.0." As explained in detail in the previous chapter, this mental model is about amplifying the narrative of stakeholder capitalism, while mostly continuing to pursue shareholder capitalism. It dominates the current state of sustainability-as-usual in business and can be seen as a "softer" version of the "original" shareholder capitalism mental model that has dominated business-as-usual for half a century. Not only does this mental model define the relationships between societal responsibilities and profit maximization considerations (with the latter having the upper hand), but it also promotes the illusion of a meaningful change (Bebchuk and Tallarita 2020).

The journey of SR has evolved tremendously over the last five decades, with developments that continued to improve the quality of the reporting in an attempt to show greater efficacy. Some things have definitely changed, as SR is no longer the exception but the rule for large companies (KPMG IMPACT 2020). It has also become more structured, strategic, and perhaps somewhat closer to financial reporting. Yet companies still produce lengthy reports that seem to have little value in terms of advancing their sustainability performance. This state of SR, where interventions generate slow and incremental improvements at best, is likely to continue as long as the deeper levels in the iceberg model are not contemplated (see Fig. 3.2). Making SR considerably more effective will require to address sustainability-as-usual and the mental model it is grounded in. As the next chapter shows, this seems to be the case not only with SR but also with the circular economy. In both cases, the mental model of shareholder capitalism 2.0 limits the capacity of companies to achieve meaningful changes and avoid the crash into the mountain Elkington (2018) was warning about in his airplane metaphor.

References

Al-Homery, Hussein Abdou, Hasbullah Ashari, and Azizah Ahmad. 2019. "The Application of System Thinking for Firm Supply Chain Sustainability: The Conceptual Study of the Development of the Iceberg Problem Solving Tool (IPST)." *International Journal of Supply Chain Management* 8: 951–56.

Baue, Bill. 2014. "#SustyGoals 2: A Dialogue with Allen White of GISR, the Godfather of Sustainability Context (Part 2)." *Sustainable Brands*. https://rb.gy/nplcsi.

————. 2019. "Compared to What? A Three-Tiered Typology of Sustainable Development Performance Indicators From Incremental to Contextual to Transformational." 2019–5. Geneva. https://www.unrisd.org/baue.

Baue, Bill, and Ralph Thurn. 2020. "The Need for Corporate Transformation in an Era of System Change." *Medium*, January 22. https://rb.gy/wohltr.

Bebchuk, Lucian A., and Roberto Tallarita. 2020. "The Illusory Promise of Stakeholder Governance." *Cornell Law Review* 106: 91–178. https://doi.org/10.2139/ssrn.3544978.

Bednárová, Michaela, Roman Klimko, and Eva Rievajová. 2019. "From Environmental Reporting to Environmental Performance." *Sustainability* 11 (9): 1–12.

Belkhir, Lotfi, Sneha Bernard, and Samih Abdelgadir. 2017. "Does GRI Reporting Impact Environmental Sustainability? A Cross-Industry Analysis of CO_2 Emissions Performance Between GRI-Reporting and Non-reporting Companies." *Management of Environmental Quality: An International Journal* 28 (2): 138–55. https://doi.org/10.1108/MEQ-10-2015-0191.

Bernow, Sara, Jonathan Godsall, Bryce Klempner, and Charlotte Merten. 2019. "Refining Sustainability Reporting for Investors." *McKinsey*, August 7. https://rb.gy/du5uvx.

Bhattacharya, C. B. 2020. *Small Actions, Big Difference: Leveraging Corporate Sustainability to Drive Business and Societal Value.* New York, NY: Routledge.

Bjørn, Anders, Niki Bey, Susse Georg, Inge Røpke, and Michael Zwicky Hauschild. 2017. "Is Earth Recognized as a Finite System in Corporate Responsibility Reporting?" *Journal of Cleaner Production* 163 (October): 106–17. https://doi.org/10.1016/j.jclepro.2015.12.095.

Blowfield, Michael, and Alan Murray. 2011. *Corporate Responsibility.* 2nd ed. Oxford: Oxford University Press.

BSR. 2012. "BSR at 20: Accelerating Progress." https://rb.gy/unew4u.

Buhr, Nola, Rob Gray, and M. J. Milne. 2014. "Histories, Rationales, Voluntary Standards and Future Prospects for Sustainability Reporting: CSR, GRI, IIRC and Beyond." In *Sustainability Accounting and Accountability*, edited by Jeffrey Unerman, Jan Bebbington, and Brendan O'Dwyer, 2nd ed., 51–70. London: Routledge.

CDP, CDSB, GRI, IIRC, and SASB. 2020. "Statement of Intent to Work Together Towards Comprehensive Corporate Reporting." https://rb.gy/lz6nuz.

Christensen, Hans Bonde, Luzi Hail, and Christian Leuz. 2019. "Adoption of CSR and Sustainability Reporting Standards: Economic Analysis and Review." *SSRN Electronic Journal*, August. https://doi.org/10.2139/ssrn.3427748.

Clayton, Anthony M. H., and Nicholas J. Radcliffe. 1996. *Sustainability: A Systems Approach.* London: Earthscan.

Cohn, Michael. 2020. "IIRC and SASB Plan Merger for Next Year." *Accounting Today*, November 25. https://rb.gy/djyike.

Deloitte. 2020. "Sustainability Reporting Strategy Creating Impact Through Transparency." https://en-rules.hkex.com.hk/node/1892.

Dentoni, Domenico, Jonatan Pinkse, and Rob Lubberink. 2020. "Linking Sustainable Business Models to Socio-ecological Resilience Through Cross-Sector Partnerships: A Complex Adaptive Systems View." *Business and Society*, June. https://doi.org/10.1177/0007650320935015.

Diaz-Rainey, Ivan. 2020. "New Zealand Will Make Big Banks, Insurers and Firms Disclose Their Climate Risk: It's Time Other Countries Did Too." *The Conversation*, September 17. https://rb.gy/hyzfr0.

Downar, Benedikt, Jürgen Ernstberger, Stefan J. Reichelstein, Sebastian Schwenen, and Aleksandar Zaklan. 2020. "The Impact of Carbon Disclosure Mandates on Emissions and Financial Operating Performance." Discussion Research Paper No. 20-038. https://doi.org/10.2139/ssrn.3693670.

Eccles, Bob. 2020. "Crunch Time: Global Standard Setters Set the Scene for Comprehensive Corporate Reporting." *Forbes*, October 8. https://rb.gy/kqfd9n.

Elkington, John. 1997. *Cannibals with Forks: The Triple Bottom Line of 21st Century Business*. Oxford, UK: Capstone Publishing.

———. 2018. "Introductory Comments." 5th International Reporting 3.0 Conference. https://www.2018.reporting3.org/.

Ernst & Ernst. 1978. "Social Responsibility Disclosure: 1978 Survey." Cleveland, OH.

Espinosa, Angela, and Terry Porter. 2011. "Sustainability, Complexity and Learning: Insights from Complex Systems Approaches." *Learning Organization* 18 (1): 54–72. https://doi.org/10.1108/09696471111096000.

European Court of Auditors. 2019. "Rapid Case Review: Reporting on Sustainability: A Stocktake of EU Institutions and Agencies." Luxembourg. https://rb.gy/qk6n23.

EY. 2014. "Sustainability Reporting-the Time Is Now." https://rb.gy/tfbyex.

EY and Boston College Center for Corporate Citizenship. 2013. "Value of Sustainability Reporting A Study by EY and Boston College Center for Corporate Citizenship." https://rb.gy/c3adb0.

Faeste, Lars, Martin Reeves, and Kevin Whitaker. 2019. "Winning the '20s: The Science of Change." *BCG Henderson Institute*, April 9. https://bcghendersoninstitute.com/winning-the-20s-the-science-of-change-95db378c5d91.

Falk, Johan. 2020. "THE 1.5°C Business Playbook." www.exponentialbusiness.org.

Fink, Larry. 2020. "Larry Fink's Letter to CEOs: A Fundamental Reshaping of Finance." *BlackRock*, January. https://rb.gy/vs0a3w.

————. 2021. "Larry Fink's 2021 Letter to CEOs." *BlackRock.* https://rb.gy/terps5.

G&A Institute. 2020. "2020 Flash Report S&P." https://rb.gy/9vmrvb.

Gaziulusoy, Idil. 2011. "Complexity and Co-evolution." *Dr. İdil Gaziulusoy—Design for Sustainability Transitions*, June 27. https://rb.gy/77vqb9.

Gaziulusoy, Idil, and Han Brezet. 2015. "Design for System Innovations and Transitions: A Conceptual Framework Integrating Insights from Sustainablity Science and Theories of System Innovations and Transitions." *Journal of Cleaner Production* 108 (December): 558–68. https://doi.org/10.1016/j.jclepro.2015.06.066.

Glauner, Friedrich. 2019. "Redefining Economics: Why Shared Value Is Not Enough." *Competitiveness Review* 29 (5): 497–514. https://doi.org/10.1108/CR-07-2016-0042.

Global Reporting Initiative. 2002. "GRI Sustainability Reporting Guidelines." Boston, MA.

Gray, Rob H. 2006. "Does Sustainability Reporting Improve Corporate Behaviour? Wrong Question? Right Time?" *Accounting and Business Research* 36 (4): 65–88.

GRI. 2011. "Sustainability Reporting." Global Reporting Initiative (GRI).

————. 2016. "Empowering Sustainable Decisions: GRI's Annual Report 2015–2016." Amsterdam.

————. 2018. "GRI Standards Glossary 2018."

Hess, David. 2014. "The Future of Sustainability Reporting as a Regulatory Mechanism." In *Law and the Transition to Business Sustainability*, edited by Daniel R. Cahoy and Jamison E. Colburn, 125–39. Cham: Springer International Publishing. https://doi.org/10.1007/978-3-319-04723-2_7.

Holland, John H. 2006. "Studying Complex Adaptive Systems." *Journal of Systems Science and Complexity* 19 (1): 1–8. https://doi.org/10.1007/s11424-006-0001-z.

IIRC. 2013. "The International <IR> Framework." https://rb.gy/l9jpzk.

Iñigo, Edurne A., and Laura Albareda. 2016. "Understanding Sustainable Innovation as a Complex Adaptive System: A Systemic Approach to the Firm." *Journal of Cleaner Production.* Elsevier Ltd. https://doi.org/10.1016/j.jclepro.2016.03.036.

Kim, Daniel H. 1999. *Introduction to Systems Thinking.* Pegasus Communications.

Kolk, Ans, Mark van der Veen, L. Wateringen, D. Veldt, and S. Walhain. 1999. "KPMG International Survey of Environmental Reporting 1999."

KPMG IMPACT. 2020. "The Time Has Come: The KPMG Survey of Sustainability Reporting 2020." https://rb.gy/wrmsmn.

Kuh, Thomas, Andre Shepley, Greg Bala, and Michael Flowers. 2020. "Dynamic Materiality: Measuring What Matters." *SSRN Electronic Journal*, February. https://doi.org/10.2139/ssrn.3521035.

de los Reyes, Gastón, and Markus Scholz. 2019. "The Limits of the Business Case for Sustainability: Don't Count on 'Creating Shared Value' to Extinguish Corporate Destruction." *Journal of Cleaner Production* 221 (June): 785–94. https://doi.org/10.1016/j.jclepro.2019.02.187.

Maas, K. E. H., and M. C. Vermeulen. 2015. "A Systemic View on the Impacts of Regulating Non-financial Reporting." Roterdam.

Mathews, M. R. 1997. "Twenty-Five Years of Social and Environmental Accounting Research: Is There a Silver Jubilee to Celebrate?" *Accounting, Auditing and Accountability Journal*. MCB UP Ltd. https://doi.org/10.1108/EUM0000000004417.

Matisoff, Daniel C. 2013. "Different Rays of Sunlight: Understanding Information Disclosure and Carbon Transparency." *Energy Policy* 55 (April): 579–92. https://doi.org/10.1016/j.enpol.2012.12.049.

McElroy, Mark W. 2008. "Social Footprints: Measuring the Social Sustainability Performance of Organizations." University of Groningen.

———. 2020. "A Brief History of the Birth of Context-Based Sustainability (CBS): A Personal Account."

McElroy, Mark W., and Jo M. L. van Engelen. 2012. *Corporate Sustainability Management: The Art and Science of Managing Non-financial Performance.* London: Earthscan.

Milne, Markus J., and Rob Gray. 2013. "W(h)ither Ecology? The Triple Bottom Line, the Global Reporting Initiative, and Corporate Sustainability Reporting." *Journal of Business Ethics* 118 (1): 13–29. https://doi.org/10.1007/s10551-012-1543-8.

Mohin, Tim. 2019. "Testimony Before the U.S. House Committee on Financial Services, Subcommittee on Investor Protection, Entrepreneurship and Capital Markets," July 10.

Monat, Jamie P., and Thomas F. Gannon. 2015. "What Is Systems Thinking? A Review of Selected Literature Plus Recommendations." *American Journal of Systems Science* 4 (1): 11–26. https://doi.org/10.5923/j.ajss.20150401.02.

Porter, Terry, and Robbin Derry. 2012. "Sustainability and Business in a Complex World." *Business and Society Review* 117 (1): 33–53. https://doi.org/10.1111/j.1467-8594.2012.00398.x.

Raworth, Kate. 2017. *Doughnut Economics: Seven Ways to Think Like a 21st-Century Economist.* London: Random House.

Reeves, Martin, and Johann D. Harnoss. 2017. "The Business of Business Is No Longer Just Business." *BCG*, June 6.

Reeves, Martin, and Simon Levin. 2017. "Think Biologically: Messy Management for a Complex World." *BCG*, July 18.

Robinson, Christine, Inna Vodovoz, Kristen Sullivan Sullivan, and Jennifer Burns. 2019. "Sustainability Disclosure Goes Mainstream." *Deloitte.*

SASB. n.d. "Standards Overview." Accessed October 10, 2020. https://www.sasb.org/standards-overview/.

———. 2018. "Materiality Map." https://www.sasb.org/standards-overview/materiality-map/.

Schneider, Anselm. 2019. "Bound to Fail? Exploring the Systemic Pathologies of CSR and Their Implications for CSR Research." *Business and Society* 59 (7): 1303–38. https://doi.org/10.1177/0007650319856616.

Scott, Louise, and Alan McGill. 2018. "From Promise to Reality: Does Business Really Care About the SDGs?" www.pwc.com/sdgreportingchallenge.

———. 2019. "Creating a Strategy for a Better World How the Sustainable Development Goals Can Provide the Framework for Business to Deliver Progress on Our Global Challenges." www.pwc.com/sdgchallenge.

Serafeim, George. 2020. "Social-Impact Efforts That Create Real Value." *Harvard Business Review* 98 (5): 38–48.

Smith, Wendy K., and Marianne W. Lewis. 2011. "Toward a Theory of Paradox: A Dynamic Equilibrium Model of Organizing." *The Academy of Management Review* 36 (2): 381–403. http://www.jstor.org.libproxy.newschool.edu/stable/41318006.

Sridhar, Kaushik, and Grant Jones. 2013. "The Three Fundamental Criticisms of the Triple Bottom Line Approach: An Empirical Study to Link Sustainability Reports in Companies Based in the Asia-Pacific Region and TBL Shortcomings." *Asian Journal of Business Ethics* 2 (1): 91–111. https://doi.org/10.1007/s13520-012-0019-3.

SustainAbility and UNEP. 1997. "Engaging Stakeholders: The 1997 Benchmark Survey: The Third International Progress Report on Company Environmental Reporting." London.

Tang, Samuel, and David Demeritt. 2018. "Climate Change and Mandatory Carbon Reporting: Impacts on Business Process and Performance." *Business Strategy and the Environment* 27 (4): 437–55. https://doi.org/10.1002/bse.1985.

TCFD. 2020. "Task Force on Climate-Related Financial Disclosures Overview." https://rb.gy/a45l2m.

The Coca Cola Company. 2020. "2019 Business and Sustainability Report." https://rb.gy/dqwe7a.

The Conference Board. 2020a. "Sustainability Practices: 2019 Edition—Executive Summary." *The Conference Board.* https://rb.gy/8twh9b.

———. 2020b. "Companies Have Significantly Increased Disclosure on Sustainability Issues, But Pressure to Do More Continues." *Sustainability-Reports.Com*, February 10. https://rb.gy/lshp0u.

The Evidence Centre. 2010. "Complex Adaptive Systems Evidence Scan." https://rb.gy/kgxx3y.

Thurm, Ralph. 2017. "Blueprint 1: Reporting a Principles-Based Approach to Reporting Serving a Green, Inclusive and Open Economy." https://rb.gy/o1y46g.

Unerman, Jeffrey, and Franco Zappettini. 2014. "Incorporating Materiality Considerations into Analyses of Absence from Sustainability Reporting." *Social and Environmental Accountability Journal* 34 (3): 172–86. https://doi.org/10.1080/0969160X.2014.965262.

Utting, Peter. 2020. "Policy Brief 28." Geneva, Switzerland. www.unrisd.org/rpb28.

WEF. 2020. "Measuring Stakeholder Capitalism Towards Common Metrics and Consistent Reporting of Sustainable Value Creation." https://rb.gy/z0h5by.

Zhang, Yue-Jun, and Jing-Yue Liu. 2020. "Overview of Research on Carbon Information Disclosure." *Frontiers of Engineering Management* 7 (1): 47–62. https://doi.org/10.1007/s42524-019-0089-1.

CHAPTER 4

The Rise of the (Mc)Circular Economy

Abstract The circular economy (CE) has become a key sustainability strategy, offering exciting pathways to convert critical environmental challenges into business opportunities that are grounded in innovation, new business models, and disruptive technologies. This chapter reviews the premise of the CE, pointing to the challenges that pursuing this strategy may have as long as it is subjected to sustainability-as-usual. Examining the opportunities and challenges of the CE on different levels (product, business model, company), the chapter focuses on the CE's proposition of decoupling growth of economic activity from the consumption of finite resources. Three main approaches to the CE and decoupling are then presented: the skeptics', the champions', and those seeing the CE as a steppingstone to more effective strategies. After reviewing the missing social dimension in the CE, the chapter moves to discuss how sustainability-as-usual impacts the CE, and why the former could lead the latter to evolve into what I describe as the "McCircular economy."

Keywords Circular economy · Decoupling · Sufficiency · Economic growth · Circularity · Circular design

© The Author(s), under exclusive license to Springer Nature Switzerland AG 2021
R. Godelnik, *Rethinking Corporate Sustainability in the Era of Climate Crisis*, https://doi.org/10.1007/978-3-030-77318-2_4

67

If there is one approach to redesigning how companies create value that almost everyone—from CEOs to sustainability experts to policymakers—is excited about, it is the circular economy (CE). The CE appears to provide a compelling pathway to a meaningful sustainable change in business, while unlocking trillions of dollars of economic activity.[1] It presents ways to convert critical environmental challenges into business opportunities that are grounded in innovation, new business models, and disruptive technologies. Another way to look at it is that the CE offers companies and policymakers a playbook for transformative changes that will generate both environmental and economic value. For consumers, the CE also provides an attractive vision: "Circularity promises an exciting world of technological progress where we can have it all – the trendiest jean silhouette, the latest gadgets, single-use plastics – without harming the planet" (Cline 2021, 33). With such a big upside and almost no downside, who cannot love the premise of the CE?

The growing urgency around the climate crisis seems to have made the case for circular solutions even stronger, driven by the notion that "the circular and low-carbon agendas are complementary and mutually supportive" (Circle Economy 2020, 12). The economic downturn following the COVID-19 pandemic has also led business leaders and policymakers to emphasize the value of CE solutions in building the economy back better (Ellen MacArthur Foundation 2020; Council of the EU 2020). Overall, it looks like circularity is gaining momentum as an instrumental approach to meaningful change on every level, from products to companies to economies. At the same time, as this chapter highlights, there is a great risk that—as promising as they may be—circular solutions will end up providing relatively little or perhaps even negative value, as long as they are subjected to the mental model of sustainability-as-usual.

THE PREMISE OF THE CE: OPPORTUNITIES AND CHALLENGES

Interest in the CE has been growing over the last three decades. The introduction of the CE as a concept is attributed to Pearce and Turner (1989), and its roots are connected to earlier work by Boulding

---

[1] Lacy and Rutquist (2015) estimated the economic potential of the CE at $4.5 trillion by 2030. In a more recent publication, Lacy et al. (2020) suggested that this estimate is still valid.

(1966), Stahel (Stahel and Reday 1976; Stahel 1982), and others (see Ghisellini et al. 2016). The concept of the CE is also drawing on different schools of thought, including industrial ecology, regenerative design, biomimicry, and cradle to cradle (Ellen MacArthur Foundation 2013). While frequently associated with improving waste management (Ghisellini et al. 2016), at its core the CE offers an alternative approach to value creation that aims to replace the current "take–make–waste" linear economic system. According to the Ellen MacArthur Foundation (EMF) (2019), the CE is built on three principles: designing out waste and pollution, keeping products and materials in use, and regenerating natural systems.[2] Furthermore, EMF, which is currently the most prominent promoter of the CE, sees it as a way to decouple economic activity (or growth) from the consumption of finite resources. This is also how the EU and China view the CE, according to Ghisellini et al. (2016).

On a product-based level, the CE approach aims to significantly reduce products' environmental impact while enhancing their durability. According to Balkenende et al. (2018), "this is achieved by designing for long product life and by enabling effective repair, refurbishment, remanufacture, parts harvesting and recycling in order to loop back used products, components and materials into the economic system" (504). These strategies can also be described in terms of closing, slowing, and narrowing resource loops (Bocken et al. 2016). They have also been integrated into circular business models, defined as "business models that are cycling, extending, intensifying, and/or dematerialising material and energy loops to reduce the resource inputs into and the waste and emission leakage out of an organisational system" (Geissdoerfer et al. 2020, 8).

While a growing number of companies have been exploring how to redesign everything from raw materials to production processes to business models in a more circular manner, the effectiveness of the CE as an approach to significantly improving companies' sustainability remains unclear. On a product or business model level, it seems plausible that circular design strategies, such as designing for material efficiency, material substitution, or design for durability, may result in reduced environmental impacts (Ellen MacArthur Foundation 2019). For example, Cooper (2010) notes that "broadly speaking, doubling the life span of

---

[2] There are over 100 definitions of the CE. For a more detailed analysis of the different definitions, see Kirchherr et al. (2017).

most consumer durables will halve their environmental impact" (220). At the same time, there are more skeptical views, such as that put forward by Hickel (2020), who notes that "in the end, only a small fraction of our total material use has circular potential" (156).

Even if we assume that circular strategies can generate meaningful reductions in environmental impacts on a microlevel (i.e., at the level of the product or business model), a key challenge is whether this can be done on a macro level, be that a company, an industry, or an economy as a whole. The main debatable variable is growth in economic activity. For example, if a company reduces the carbon footprint of a product by 50% using circular design strategies, but doubles the sales of this product, then the total carbon footprint will not change at all. There seems to be an inevitable tension between the sustainability ambitions of the CE and the growth-based economy it is situated in. As De Decker (2018) explains, "Growth makes a circular economy impossible, even if all raw materials were recycled and all recycling was 100% efficient. The amount of used material that can be recycled will always be smaller than the material needed for growth. To compensate for that, we have to continuously extract more resources" (para. 15). Beyond the potential difficulties of significantly reducing the use of new raw materials in a system optimized for growth, there is also the issue of energy use. While the goal would be to use 100% renewable energy, the processes involved in slowing, narrowing, and closing the material loops also require energy. Thus, in a growth-based economy, as Hickel notes, it will be very difficult to provide the necessary energy supply for the economy using only renewables (Dubner 2020).

These concerns echo earlier questions about the potential of eco-efficiency and decoupling strategies to advance sustainability at scale (see, e.g., Jackson 2009). So far, empirical evidence appears to support those who have doubts about decoupling (see, e.g., Parrique et al. 2019); however, proponents of decoupling (e.g., UNEP 2011) have put their faith in technological innovation and supporting policies that will make decoupling work. Thus far both of these potential drivers seem to be lagging, leading to disappointing results especially when it comes to technological innovation, which is still more of a promise than a reality (Cline 2021). For example, in the apparel industry, EMF (2017) notes that "existing recycling technologies for common materials need to drastically improve their economics and output quality to capture the full value of the materials in recovered clothing" (25).

Three Approaches to the CE and Decoupling

Overall, it seems there are currently three different approaches to CE and decoupling. There is a group of skeptics who do not believe that "closing the circle" has much value when it comes to advancing sustainability if the circle keeps growing. In general, this group seems to include advocates of degrowth (e.g., Hickel 2020), who associates the CE with "green growth," which they find to be a "misguided objective" (Hickel and Kallis 2020). This group considers the CE as a strategy that could improve flows of materials and energy but that, as long as it is part of a growth-based economy, cannot create the transformational changes that are needed to address issues like the climate crisis. Parrique et al. (2019) suggest that the premise of decoupling appears to be unrealistic, while Fletcher and Rammelt (2017) call decoupling a neoliberal fantasy. Cline (2021) sees it as "a myth that we'll have to leave behind if we ever want to realize the dream of a circular economy" (33). Zink and Geyer (2017) add further critique based on economic analysis, arguing that CE activities are likely to end up increasing overall production and consumption, and this could diminish if not eliminate the benefits of the CE. They call this effect "circular economy rebound." Makov and Font Vivanco (2018), who looked to measure this rebound effect in a case study of reused smartphones sold in the U.S., found that "about one third, and potentially the entirety, of emission savings resulting from smartphone reuse could be lost due to the rebound effect" (1).

While sharing a general agreement with the skeptics' assessment that the CE could have a significant sustainable value only within a framework not subjected to economic growth, the second group provides a somewhat more nuanced approach to the CE. This group also sees the CE as an insufficient strategy on its own, but at the same time considers its potential as a steppingstone to more effective strategies that could advance sustainability more meaningfully. A key example of such an effective strategy that could "stand on the shoulders" of the CE is sufficiency, which is defined as having enough to live well with no unnecessary excess (Bocken and Short 2020). A sufficiency-based approach, according to Bocken and Short (2016), is more "consumption-focused, aimed at moderating end-user consumption: encouraging consumers to make do with less" (42). A key difference between this and the dominant CE approach (the third group) is that instead of asking how to improve product recycling or enhance the product life, this group asks if the consumer needs the product in the first place.

The third group ("CE champions"), which is currently the most dominant among the three, sees no issue with the framing of the CE as a decoupling mechanism, being driven by the belief that living within our ecological means and economic growth are not mutually exclusive. This approach is supported by business leaders and policymakers alike, who view the CE as an effective way to untie the Gordian knot of unsustainable economic activity. They believe the CE provides companies and economies with a clear pathway to fight climate change, while continuing to grow but in a sustainable way. Furthermore, these CE champions like to present circularity as an exciting new narrative that reframes sustainability in terms of innovation. As Dame Ellen MacArthur, the founder of the EMF, explained in an interview, "There's a different narrative. There's a different story. There's a story around innovation, creatively decoupling growth and resource constraints. We can do this so much better" (Phipps 2019, para. 15). In addition to being an innovation driver, the CE is portrayed by its champions as a resilience builder (Ellen MacArthur Foundation 2020), which makes it seem an even more compelling proposition, especially for policymakers in the aftermath of COVID-19. Altogether, this group sees the CE as the ultimate "win–win" strategy to lead companies and economies toward a sustainable future.

The interest of policymakers from the third group in the potential of circularity (e.g., the new EU Circular Economy Action Plan)[3] echoes decades of sustainable development agenda, which sought to balance economic growth with sustainability goals. The UN's Sustainable Development Goals (SDGs),[4] the latest manifestation of this agenda, represent a good example of this approach. Combining social and environmental goals for 2030, the 17 SDGs (and the 169 targets they include) aim among other things to "promote sustained economic growth, higher levels of productivity and technological innovation" (UNDP 2021), aims which seem to be well-aligned with the CE narrative. It is therefore not too surprising to hear warm endorsements of the CE from UN officials, who suggest, for example, that "circularity and sustainable consumption and production are essential to delivering on every single multilateral

---

[3] See https://ec.europa.eu/commission/presscorner/detail/en/ip_20_420.

[4] See https://www.un.org/development/desa/dspd/2030agenda-sdgs.html.

agreement, from the Sustainable Development Goals (SDGs) to the Paris Agreement and beyond" (Andersen 2020, para. 15). Overall, applying CE strategies to support a sustainable growth agenda seems to be increasingly endorsed by the UN and government bodies worldwide, who believe that sustainability should go hand in hand with what Jackson (2017) describes as "a growth-jobs-prosperity trifecta." This connection between the SDGs and the CE is also supported by research suggesting that "CE practices, potentially, can contribute directly to achieving a significant number of SDG targets" (Schroeder et al. 2019, 77).

The Missing Social Dimension

The social dimension, or to be more precise the lack of it, has long been considered the Achilles heel of the CE. Evolving mainly as a response to the environmental impacts of the production and consumption systems, the CE has been, as Murray et al. (2017) describe it, "virtually silent on the social dimension, concentrating on the redesign of manufacturing and service systems to benefit the bio-sphere" (376). Although it is not completely absent from the CE narrative,[5] the social dimension of the CE is usually relatively marginal and certainly does not carry the same weight as it does in the context of sustainability, which aims to balance environmental, social, and economic considerations (Merli et al. 2018).

The CE's limited consideration of social issues has become more noticeable lately with the growing attention that is being paid to social justice issues. While there is a debate about the overall ability of CE strategies to provide an effective response to critical issues such as the climate crisis,[6] there is general agreement that the CE does not approach these issues through a social lens. Perhaps as a reflection of its industrial context, the CE's approach is more material-centered than human-centered, focusing first and foremost on improving material flows in the hope that these changes will generate societal benefits for all (see, e.g., Ellen MacArthur Foundation 2019). This "social deficiency" may be compensated for by combining the CE with other strategies that prioritize social considerations (e.g., the EU Green Deal). However, there is

---

[5] See, for example, the EMF's vision for a CE for plastic at https://www.newplasticseconomy.org/assets/doc/npec-vision.pdf.

[6] Seem for example, the EMF (2019) vs. Hickel and Kallis (2020).

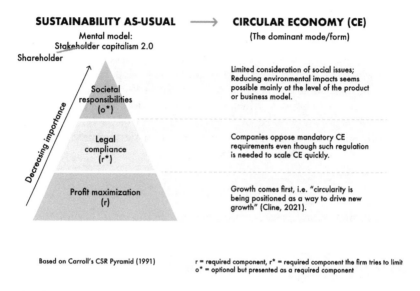

Fig. 4.1   How sustainability-as-usual is reflected in the circular economy

the risk that, by ignoring the growing recognition that environmental and social issues cannot be separated since they are strongly connected to one another, advocating for a focus on only one dimension will make the CE into a much less valuable proposition.

The CE as a Symbol of Sustainability-as-Usual

To a great extent, the CE has become one of the poster children of the state of sustainability-as-usual, which was described in detail in Chapter 2. The CE appears to check all the sustainability-as-usual boxes, offering to create a positive environmental impact, but at the same time also grounded in the mental model of shareholder capitalism 2.0, in which economic considerations transcend those of sustainability (see Fig. 4.1).[7] With profit maximization (or growth) as a top priority, it is not clear what chance the CE has of driving transformational change in a relatively short

[7] This characterization refers to the dominant form of CE, i.e., the one championed by the aforementioned third group believes in the ability of the CE to decouple economic growth from resource consumption.

time, which is what is needed to address critical sustainability issues, most urgent among these being the climate crisis.

Feeling more pressure to meet bottom-line expectations than to significantly reduce their environmental impacts, companies are likely to move forward slowly with new circular initiatives, especially when the profitability of these initiatives is unclear.[8] Furthermore, the prioritization of profits over sustainability drives companies to consider circular practices as additional revenue streams rather than as replacements for their non-circular practices (Cline 2021). An indication of these trends can be found in the annual circular gap reports, which provide an estimate of the level of circularity in the world. According to these reports, circularity was 9.1% in 2018, while two years later it actually fell to 8.6%[9] (Circle Economy 2021).

One key sustainability-as-usual element that is visible in the CE is how regulation is viewed. While it has become clear that implementation of the CE at scale requires legislation and policy support (see, e.g., Ellen MacArthur Foundation 2017; Circle Economy 2021), companies seem to prefer that their responsibilities to align with the CE will remain voluntary. One example is the legislative efforts in support of the "right to repair," which aims to ensure companies "make their parts, tools and information available to consumers and repair shops in order to keep devices from ending up in the scrap heap" (Rosa-Aquino 2020, para. 7). Companies seem reluctant to support such legislative efforts (Gault 2020; Proctor 2019), even though these efforts aim to help extend the lifetime of products, one of the key tenets of the CE.

Interestingly, companies appear to be in favor of incentives or subsidies that support the sale of products that apply CE principles (e.g., use of recycled materials), but they change their tone once the legislative efforts are directed at establishing new "rules of the game" that mandate rather than just ask for changes in product design, lifetime, duty of care, and so on. While it is doubtful that without such regulation in place circularity will become a meaningful sustainability strategy (Maitre-Ekern et al. 2020), it appears that the mental model of shareholder capitalism 2.0 is

---

[8] A global survey of Bureau Veritas (2020) found that lack of evidence of profitability and cheap virgin materials (another profitability consideration) were two of the three top barriers to adoption of CE practices.

[9] According to the report's authors, global circularity needs to reach 17% to enable a reduction in emissions in accordance with the Paris Agreement.

more powerful than any other consideration. Overall, just as with sustainability reporting or setting carbon reduction goals, companies seem to have a clear preference to keep the current voluntary framework in place, so that they have control over the timeline, scope, and narrative of the transition to a CE.

The Mc(Circular) Economy?

I can see many similarities between the CE in its current dominant form (i.e., the one championed by the aforementioned third group) and mindfulness as portrayed in Purser's (2019) book *McMindfulness: How Mindfulness Became the New Capitalist Spirituality*. In his critique of the popularized form of mindfulness, Purser suggests that "mindfulness is hostage to the neoliberal mindset…This prevents it being offered as a tool of resistance, restricting it instead to a technique for 'selfcare'" (20). He points out that "anything that offers success in our unjust society without trying to change it is not revolutionary – it just helps people cope. However, it could also be making things worse. Instead of encouraging radical action, it says the causes of suffering are disproportionately inside us, not in the political and economic frameworks that shape how we live" (7–8). I find it unfortunate that this critique could be easily applied to the CE, which seems to show similar symptoms, focusing on operational challenges as the core issues while paying little to no attention to the ills of the economic system.

While the CE may not have had radical ambitions in the first place, this does not mean that it should evolve into the "McCircular economy," keeping companies' focus on "doing more with less" without the required consideration of the framework in which these companies operate. Nevertheless, it looks as if this will be the situation as long as the CE is subjected to sustainability-as-usual, which, to paraphrase Purser, makes the CE a hostage of the shareholder capitalism 2.0 mindset. In that sense, the growing popularity of the CE in its dominant form may be a warning sign rather than a sign of progress. If indeed Fletcher and Tham (2019) are right that "the circular economy has gained traction and substantial interest perhaps because it aligns with existing commercial practices, suggesting that business-(almost)-as-usual is possible" (20), then the growing popularity of the CE could be a step backward, not forward.

Overall, the form and shape of the CE, just like that of mindfulness, is determined by the dominant mental model. As long as this is the mental model of sustainability-as-usual (i.e., shareholder capitalism 2.0), the CE may have a hard time evolving beyond the rather incremental "weak" approach to sustainability (Hobson and Lynch 2016) it represents now. While the "nearly revolutionary rhetoric of its advocates" (Corvellec et al. 2020, 98) may provide a different impression of the CE, the single-digit percentage of circularity in the world (Circle Economy 2020, 2021), appears to support the grim perspective of Hobson and Lynch. The state of the CE is far from static and it will certainly change, but if we keep focusing on improving the CE strategies without changing the state of sustainability-as-usual, we will probably end up with one form or another of McCircular Economy. If we want a different outcome for the CE we need to figure out first how to move away from sustainability-as-usual. This quest will be explored in the next chapters of the book.

REFERENCES

Andersen, Inger. 2020. "Scaling Up Circularity Is Vital to Sustainability." *UNEP*, September 29. https://rb.gy/dsyblq.

Balkenende, Ruud, Nancy M. P. Bocken, and Conny Bakker. 2018. "Design for the Circular Economy." In *Routledge Handbook of Sustainable Design*, 498–513. Routledge. https://doi.org/10.4324/9781315625508-42.

Bocken, Nancy M. P., Ingrid de Pauw, Conny Bakker, and Bram van der Grinten. 2016. "Product Design and Business Model Strategies for a Circular Economy." *Journal of Industrial and Production Engineering* 33 (5): 308–20. https://doi.org/10.1080/21681015.2016.1172124.

Bocken, Nancy M. P., and Samuel W. Short. 2016. "Towards a Sufficiency-Driven Business Model: Experiences and Opportunities." *Environmental Innovation and Societal Transitions* 18: 41–61. https://doi.org/10.1016/j.eist.2015.07.010.

———. 2020. "Transforming Business Models: Towards a Sufficiency-Based Circular Economy." In *Handbook of the Circular Economy*, edited by Miguel Brandão, David Lazarevic, and Göran Finnveden, 250–65. Edward Elgar Publishing.

Boulding, Kenneth E. 1966. "The Economics of the Coming Spaceship Earth." In *Environmental Quality in a Growing Economy*, edited by Henry Jarrett, 3–14. Baltimore, MD: Resources for the Future/Johns Hopkins University Press.

Bureau Veritas. 2020. "How Circular Is the Business World Today?—Results of the Global Survey." *Bureau Veritas*, June 8. https://rb.gy/2j6eex.

Circle Economy. 2020. "The Circularity Gap Report 2020." Amsterdam.
———. 2021. "The Circularity Gap Report 2021." Amsterdam.
Cline, Elizabeth L. 2021. "Will the Circular Economy Save the Planet?" *Sierra* January/February, 28–33.
Cooper, Tim. 2010. "Policies for Longevity." In *Longer Lasting Products: Alternatives to the Throwaway Society*, edited by Tim Cooper. Farnham, UK: Gower Publishing.
Corvellec, Hervé, Steffen Böhm, Alison Stowell, and Francisco Valenzuela. 2020. "Introduction to the Special Issue on the Contested Realities of the Circular Economy." *Culture and Organization* 26 (2): 97–102. https://doi.org/10.1080/14759551.2020.1717733.
Council of the EU. 2020. "Draft Council Conclusions on Making the Recovery Circular and Green—Approval." Brussels, Belgium.
De Decker, Kris. 2018. "How Circular Is the Circular Economy?" *Low-Tech Magazine*, 1–10.
Dubner, Stephen J. 2020. "Is Economic Growth the Wrong Goal? (Ep. 429)." *Freakonomics Radio.* https://rb.gy/0qpxk5.
Ellen MacArthur Foundation. 2013. "Towards the Circular Economy: Economic and Business Rationale for an Accelerated Transition."
———. 2017. "A New Textiles Economy: Redesigning Fashion's Future."
———. 2019. "Completing the Picture: How the Circular Economy Tackles Climate Change."
———. 2020. "A Solution to Build Back Better: The Circular Economy." *Financial Times*, June 13. https://rb.gy/xqss1e.
Fletcher, Kate, and Mathilda Tham. 2019. "Earth Logic Fashion Action Research Plan." London.
Fletcher, Robert, and Crelis Rammelt. 2017. "Decoupling: A Key Fantasy of the Post-2015 Sustainable Development Agenda." *Globalizations* 14 (3): 450–67. https://doi.org/10.1080/14747731.2016.1263077.
Gault, Matthew. 2020. "Auto Industry Has Spent $25 Million Lobbying Against Right to Repair Ballot Measure." *Vice*, September 29. https://rb.gy/t4hioy.
Geissdoerfer, Martin, Marina P. P. Pieroni, Daniela C. A. Pigosso, and Khaled Soufani. 2020. "Circular Business Models: A Review." *Journal of Cleaner Production.* https://doi.org/10.1016/j.jclepro.2020.123741.
Ghisellini, Patrizia, Catia Cialani, and Sergio Ulgiati. 2016. "A Review on Circular Economy: The Expected Transition to a Balanced Interplay of Environmental and Economic Systems." *Journal of Cleaner Production* 114: 11–32. https://doi.org/10.1016/j.jclepro.2015.09.007.
Hickel, Jason. 2020. *Less Is More: How Degrowth Will Save the World.* London: William Heinemann.

Hickel, Jason, and Giorgos Kallis. 2020. "Is Green Growth Possible?" *New Political Economy* 25 (4): 469–86. https://doi.org/10.1080/13563467.2019.1598964.

Hobson, Kersty, and Nicholas Lynch. 2016. "Diversifying and De-growing the Circular Economy: Radical Social Transformation in a Resource-Scarce World." *Futures* 82 (September): 15–25. https://doi.org/10.1016/j.futures.2016.05.012.

Jackson, Tim. 2009. *Prosperity Without Growth: Economics for a Finite Planet.* London, UK: Earthscan.

———. 2017. "How to Kick the Growth Addiction." *Great Transition Initiative*, April. https://rb.gy/cutbjh.

Kirchherr, Julian, Denise Reike, and Marko Hekkert. 2017. "Conceptualizing the Circular Economy: An Analysis of 114 Definitions." *Resources, Conservation and Recycling.* Elsevier B.V. https://doi.org/10.1016/j.resconrec.2017.09.005.

Lacy, Peter, and Jakob Rutqvist. 2015. *Waste to Wealth: The Circular Economy Advantage.* London, UK: Springer. https://www.amazon.com/Waste-Wealth-Circular-Economy-Advantage/dp/1137530685.

Lacy, Peter, Jessica Long, and Wesley Spindler. 2020. *The Circular Economy Handbook: Realizing the Circular Advantage.* London, UK: Palgrave Macmillan.

Maitre-Ekern, Eléonore, Mark B. Taylor, and Maja van der Velden. 2020. "Towards a Sustainable Circular Economy: SMART Reform Proposals," 2020–12. https://papers.ssrn.com/sol3/papers.cfm?abstract_id=3596076.

Makov, Tamar, and David Font Vivanco. 2018. "Does the Circular Economy Grow the Pie? The Case of Rebound Effects From Smartphone Reuse." *Frontiers in Energy Research* 6: 39. https://doi.org/10.3389/fenrg.2018.00039.

Merli, Roberto, Michele Preziosi, and Alessia Acampora. 2018. "How Do Scholars Approach the Circular Economy? A Systematic Literature Review." *Journal of Cleaner Production.* https://doi.org/10.1016/j.jclepro.2017.12.112.

Murray, Alan, Keith Skene, and Kathryn Haynes. 2017. "The Circular Economy: An Interdisciplinary Exploration of the Concept and Application in a Global Context." *Journal of Business Ethics* 140 (3): 369–80. https://doi.org/10.1007/s10551-015-2693-2.

Parrique, T., J. Barth, F. Briens, C. Kerschner, A. Kraus-Polk, A. Kuokkanen, and J. H. Spangenberg. 2019. "Decoupling Debunked: Evidence and Arguments Against Green Growth as a Sole Strategy for Sustainability."

Pearce, David W., and R. Kerry Turner. 1989. *Economics of Natural Resources and the Environment.* Baltimore, MD: John Hopkins University Press.

Phipps, Lauren. 2019. "A Conversation with Dame Ellen MacArthur." *GreenBiz*, April 26. https://rb.gy/krdigt.

Proctor, Nathan. 2019. "Here's How Manufacturers Argue Against Repair." *U.S. PIRG*, July 1. https://rb.gy/t4vw4a.

Purser, Ronald E. 2019. *McMindfulness: How Mindfulness Became the New Capitalist Spirituality*. London: Repeater books.

Rosa-Aquino, Paola. 2020. "Fix, or Toss? The 'Right to Repair' Movement Gains Ground." *The New York Times*, October 24. https://rb.gy/gnlhve.

Schroeder, Patrick, Kartika Anggraeni, and Uwe Weber. 2019. "The Relevance of Circular Economy Practices to the Sustainable Development Goals." *Journal of Industrial Ecology* 23 (1): 77–95. https://doi.org/10.1111/jiec.12732.

Stahel, W. R. 1982. "The Product Life Factor." In *An Inquiry into the Nature of Sustainable Societies: The Role of the Private Sector*, edited by G. S. Orr, 72–105. Houston: Houston Area Research Centre.

Stahel, W. R., and G. Reday. 1976. "The Potential for Substituting Manpower for Energy, Report to the Commission of the European Communities." Brussels.

UNDP. 2021. "Goal 8: Decent Work and Economic Growth." https://rb.gy/stn9fq.

UNEP. 2011. "Decoupling Natural Resource Use and Environmental Impacts from Economic Growth." Nairobi.

Zink, Trevor, and Roland Geyer. 2017. "Circular Economy Rebound." *Journal of Industrial Ecology* 21 (3): 593–602. https://doi.org/10.1111/jiec.12545.

CHAPTER 5

The Transformation Journey, or: Why Now?

Abstract Opening the second part of the book, which focuses on replacing sustainability-as-usual with a new, desired mode, this chapter examines the question of "why now?" contemplating why this point in time (the early 2020s) could be an inflection point for corporate sustainability. Looking at the broader environment in which companies operate, this chapter lays out key economic, social, and environmental trends and events that have led to this point, with a particular focus on the multiple crises that took place in 2020 and potentially made it a pivotal year. I then consider these trends and events as part of a transformation trajectory, using Moore et al.'s four-phase process of a deliberate transformation in a social-ecological system. Based on this framework, I make the case that we are already in the early stages of a transformative change, which is driven by a convergence of social and economic pressures, the COVID-19 pandemic, and the climate crisis. Looking forward, three potential pathways for transformation are presented: fixing, reforming, or revolutionizing the system.

Keywords Transformation · Green New Deal · Climate change · COVID-19 · Capitalocene · Climate activism

© The Author(s), under exclusive license to Springer Nature
Switzerland AG 2021
R. Godelnik, *Rethinking Corporate Sustainability in the Era
of Climate Crisis*, https://doi.org/10.1007/978-3-030-77318-2_5

Right here, right now is where we draw the line. The world is waking up. And change is coming, whether you like it or not. (Greta Thunberg, U.N. 2019 Climate Action Summit[1])

After exploring the state of sustainability-as-usual and why we need to move away from it in the first part of the book, the second part will focus on what a desired change should look like, and perhaps most importantly, how to make it happen. Before delving into the "what" and "how" questions, this chapter will consider the question of timing, focusing on this point in time (the early 2020s) and on what makes it pivotal for companies in the context of corporate sustainability.

Our starting point is that this decade is critical for our future. Karen O'Brien (2020) deems it "a decisive period in human history, a time when our actions can have profound consequences for life on earth for millennia to come" (2–3). Figueres and Rivett-Carnac (2020) suggest that "it is no exaggeration to say that what we do regarding emissions reductions between now and 2030 will determine the quality of human life on this planet for hundreds of years to come, if not more" (xxii). Indeed, there is a growing recognition that the 2020s will play a key role in shaping the future, as the window of opportunity to limit global warming to 1.5 °C above pre-industrial levels is closing fast (IPCC 2018a), not to mention the risks of exceeding climate tipping points (Lenton et al. 2019).

While no one can know for sure what this decade will look like, this chapter makes the case that we are already in the early stages of a transformative change. It portrays the intersection of a number of trends and events that in the last few years have ignited a process of transformation mainly in relation to climate change. As a transformative moment (Hall et al. 2020; The Club of Rome 2020), the COVID-19 pandemic can potentially accelerate this process and make this decade into a disruptive period for companies all over the world.

To explore this point in time, I use a framework that outlines a four-phase process of a deliberate transformation in a social-ecological system (SES). Developed by Moore et al. (2014), it helps contextualize where we stand and shows what makes this moment distinct.

---

[1] Thunberg's full speech is available at https://rb.gy/tctisc.

FRAMING TRANSFORMATION

First, we should clarify what we mean by transformation. In exploring different definitions of the concept, Nalau and Handmer (2015) found that most of them consider it as a process or an act that produces a significant change from the status quo, usually a positive one. The Intergovernmental Panel on Climate Change (IPCC) (2012) defines transformation as "the altering of fundamental attributes of a system" (564). The change transformation produces can be physical or qualitative (O'Brien and Sygna 2013), as well as deliberate or forced (when a system becomes unviable) (Nalau and Handmer 2015). Lastly, the difference between a transformational change and an incremental one, according to Park et al. (2012), "lies in the extent of change, in practice manifesting in either in the maintenance of an incumbent system or process, or in the creation of a fundamentally new system or process" (119).

A deliberate transformation is considered a more appropriate response to climate change—for example, in comparison with adaptation strategies. Leichenko and O'Brien (2019) explain that while adaptation is necessary in terms of preparing ourselves to live with climate change impacts, it remains an insufficient response to climate change risks. Even with the growing understanding that limiting global warming to 1.5 °C will require significant transformations, their nature (i.e., what will be transformed, by whom, and how) is still not completely clear (Leichenko and O'Brien 2019). One framework that could help shed light on the process of transformation was developed by Moore et al. (2014), who outlined four phases of a deliberate transformative change in an SES.[2]

Moore et al. (2014) suggest that such a transformation "(1) can be triggered by a deliberate change in the key elements of either the social or ecological parts of the system across more than one scale, (2) that this change has impacts on the current dominant social-ecological feedbacks, and (3) that this leads to further changes in the structure of both the social and ecological parts of the system" (3). Knuth (2019) adds that power, meaning (or narrative), leadership, and networks are

---

[2] Within SESs "the social refers to the human dimension in its diverse facets, including the economic, political, technological, and cultural, and the ecological to the thin layer of planet Earth where there is life, the biosphere" (Folke et al. 2016). Using a SES transformation framework in the context of the corporate world, which is the context of this book, emphasizes the notion of companies (and human activity in general) as an integrated part of the biosphere.

key elements in the process of deliberate transformation. Furthermore, O'Brien (2012) points out that this process is likely to be resisted and contested by different actors who benefit from the current structures. Thus, those championing a deliberate transformation need to figure out how to overcome such barriers.

Looking at the way in which transformations materialize, Olsson et al. (2004) describe three phases that take place during them: (1) preparing for change; (2) navigating the change; and (3) building the resilience of the new trajectory of development. Based on this framework and on further research on social innovation, transition management, and social movements, Moore et al. (2014) developed the following four-phase framework for transformation processes:

1. *Triggers, or "pre-transformation."* The first phase is usually "characterized by major social or ecological disruptions, which in turn, create windows of opportunity" (M. L. Moore et al. 2014, 4) for transformation.
2. *Preparing for change.* This phase includes three processes: sense-making (of the situation), envisioning (creating an alternative vision), and gathering momentum (mobilizing support for the new direction and experimenting with new ideas).
3. *Navigating the transition.* This stage involves selecting the change processes or innovations to be prioritized, learning from earlier experiments, and the adoption (diffusion) of new ideas/innovations.
4. *Institutionalizing the new trajectory.* The final phase is about building the resilience of the new trajectory. The key processes at this stage are routinization, strengthening cross-scale relationships, and stabilization.

I will use this four-phase framework to try to pinpoint where we are on the transformation trajectory. I refer first to where I believe we were just before 2020 and to what may have been changed in the pre-2020 trajectory by the COVID-19 pandemic and the subsequent economic downturn, as well as by the protests following the murder of George Floyd. The goal of this account is not to provide a full historical overview but to focus on recent key trends and events, which could evolve into a deliberate transformation that will fundamentally change corporate sustainability in this decade.

Pre-2020

Before 2020, the world had already been struggling with various critical economic, social, and environmental issues. Global economic activity worldwide seemed to recover more or less from the severe economic crisis in 2007–2009 (also known as the Great Recession), with developed economies exhibiting a relatively weaker recovery compared to emerging markets and developing countries (Chen, Mrkaic, and Nabar 2019). Yet the levels of inequality were rising, especially in rich countries like the U.S., which in 2018 showed the highest level ever recorded of income inequality (Heeb 2019). As Nobel Laureate Economist Agnus Deaton commented in 2019, "there is this feeling that contemporary capitalism is not working for everybody" (Partington 2019, para. 3).

According to a report of the International Monetary Fund (IMF), the slow recovery experienced by many people and the growth of inequality "left tens of millions of people behind with fading hope of climbing up the social ladder" (Sedik and Xu 2020, 4). As a result, growing frustrations from the economic system led to worldwide protests in 2019, from France and Greece in Europe to Colombia and Bolivia in South America. The IMF report suggests that although there may have been different triggers to these protests, "a common theme underlying the social discontent is reported to be stagnating living standards and inequality" (Sedik and Xu 2020, 4). In the U.S., additional dimensions to this social discontent were a growing racial wealth divide (Collins et al. 2019) and the rise of social movements like Black Lives Matter,[3] which were fighting to tackle systemic racism and violence against Black people.

Besides the growing economic and social challenges, the world had been dealing with runaway climate change as well. Freedman et al. (2020) note that "the 2010s were the decade of climate change consequences— when the clear signal of human-driven extreme events fully emerged and affected the lives of millions worldwide" (para. 1). Among these events were permafrost degradation in the Arctic and the growing risk of the Amazon rainforests transitioning into Savannah. As these authors point out, however, "the biggest disruptive impacts of climate change in people's lives has come through extreme events, from heatwaves to floods and wildfires" (para. 2). A 2018 report of the IPCC on the impacts of global warming of 1.5 °C (IPCC 2018a) made it clear that we already

---

[3] See https://blacklivesmatter.com/.

see the consequences of 1 °C warming "through more extreme weather, rising sea levels and diminishing Arctic sea ice, among other changes" (IPCC 2018b, para. 6). In addition, the report highlighted the risks of warming beyond 1.5 °C, making it clear that "every extra bit ... matters" (IPCC 2018b, para. 8).

Perhaps the most notable part about the 2018 IPCC report was the explicit timeline it offered for the action necessary to limit global warming to 1.5 °C: "global net human-caused emissions of carbon dioxide (CO_2) would need to fall by about 45 percent from 2010 levels by 2030, reaching 'net zero' around 2050" (IPCC 2018a, 14). More than a wake-up call, this timeline was an important reminder of the current insufficient global response to climate change. A good example of the latter was the implementation of the Paris Agreement. This leading global framework to fight climate change was signed by most countries in 2015, with the aim of limiting "the increase in the global average temperature to well below 2 °C above pre-industrial levels and pursuing efforts to limit the temperature increase to 1.5 °C above pre-industrial levels" (UNFCCC 2015, 3). While the goals of the Paris Agreement are aimed at generating a strong response to the climate challenge, in reality, the commitments undertaken by its signatories are not even close to achieving them (UNEP 2020), and it remains unclear to what extent they could be further expanded, as planned in the agreement.[4]

The gap between the need to act with greater urgency on climate change and the generally inadequate response of governments and the private sector has led to a rise in climate activism worldwide (e.g., Extinction Rebellion,[5] the Sunrise Movement,[6] Fridays for Future [FFF]).[7] Greta Thunberg, the young Swedish climate activist whose weekly protest in front of the Swedish Parliament inspired a global movement of school students protesting against climate inaction, became the voice of young people demanding that adults "act as you would in a crisis.... as if our house is on fire. Because it is" (Thunberg 2019, para. 18). Almost as if to prove her point, climate change seemed to be accelerating toward the

---

[4] The Paris agreement includes a "ratchet mechanism" to increase the signatories' commitments over time.

[5] See https://rebellion.global/.

[6] See https://www.sunrisemovement.org/.

[7] See https://fridaysforfuture.org/.

end of the 2010s with devastating impacts, from heat and cold waves to wildfires, severe storms, and droughts (WMO 2020).

These economic, social, and environmental/climate elements were all connected. For example, the aforementioned economic and social trends contributed to the rise of populism, which in turn influenced the climate agenda (Pastor and Veronesi 2020; Weitzman 2020). On the one hand, right-wing populism has been opposing climate policies, as it perceives climate change as a cosmopolitan issue that is part of a liberal agenda (Lockwood 2018; Gardiner 2019). On the other hand, left-wing populism, which also emerged in Europe and the U.S. after the Great Recession, has been advocating for strong responses to climate change such as the Green New Deal (GND; Dyer-Witheford 2020). Also showing populist features are climate movements such as the aforementioned Extinction Rebellion and Fridays for Future (Arias-Maldonado 2020), which could be considered examples of environmental populism (Beeson 2019).

2020: A Pivotal Year

With multiple crises taking place all at once, including the COVID-19 pandemic and the resulting economic downturn, unrest over racial injustice, and the continuing effects of climate change, 2020 felt like a unique moment in history. Manifestly, it was the COVID-19 pandemic that rendered 2020 a year like no other. Considered a serious exogenous shock (Verbeke 2020), COVID-19 represents three separate types of shocks to the system—demand, supply, and financial (Triggs and Kharas 2020). This combination caused a deep recession worldwide, and while some countries did better than others, none escaped the disruption caused by the pandemic (McKinsey and Company 2020). The World Health Organization and other global entities point to the human suffering it engendered: "The COVID-19 pandemic has led to a dramatic loss of human life worldwide and presents an unprecedented challenge to public health, food systems and the world of work" (ILO, FAO, IFAD, and WHO 2020, para. 1). Furthermore, the UN Division for Inclusive Social Development (DISD; 2020) suggests that COVID-19 should be considered a human, social, and economic crisis, as it attacks "societies at their core."

A second crisis erupted after the murder of George Floyd by police officers in Minneapolis, Minnesota. In May 2020, Floyd, a 46-year old Black man, was killed during an arrest for allegedly using a counterfeit $20 bill. Floyd repeatedly pleaded "I can't breathe" while a police officer pressed his knee into Floyd's neck for more than eight minutes. These horrific images, which were captured on video, were a reminder of a long history of police brutality against Black people in the U.S., and they led to both national and international protests. Associated with the Black Lives Matter (BLM) movement, the demonstrations drew massive participation from millions of people in the U.S. alone, making it "the largest movement in the country's history" according to experts (Buchanan, Bui, and Patel 2020, para. 3). Continuing for months, these protests led to "changes in symbols of racism—from flags to statues—in stances of corporations, in hopes of real police reform, and of overdue reparations to Black Americans" (Blankenship and Reeves 2020, para. 2). Perhaps the primary change was the growing recognition of systemic racism in American society (Worland 2020), which was driven both by the protests and by the growing toll of the pandemic on minority communities.

In addition to the BLM protests and the pandemic's social, economic, and human toll, climate change was another critical element in 2020. Its impact could be seen in extreme weather events ranging from record-breaking wildfires in California and other regions in the world to an unprecedented number of named storms (Irfan 2020). Another important development in 2020 was the fiscal recovery plans announced by countries worldwide in response to the economic impacts of the COVID-19 pandemic. Many of these plans included "green" measures that create jobs and economic opportunities, while supporting the transition to a net-zero carbon economy.[8] The green recovery approach was led by the European Union and some other countries, such as the U.K., South Korea, and Japan. One factor that could help boost the momentum around the green recovery is the election of Joe Biden as President of the U.S., which in itself is one of the key events of 2020.

Evaluating the aforementioned events and trends (i.e., elements) through Moore et al.'s (2014) framework, I consider first which of these elements can be regarded as a trigger (phase 1). Based on the researchers'

---

[8] See a full list of green recovery commitments worldwide at https://www.carbonbrief. org/coronavirus-tracking-how-the-worlds-green-recovery-plans-aim-to-cut-emissions.

definitions, I consider an element to be a trigger if it is either an exogenous shock or an attempt to "intentionally disrupt a dominant state that has become rigid but which locks the system into an unsustainable trajectory" that could "weaken one or more of the social elements" (M. L. Moore et al. 2014, 3). The researchers point out that weakening social elements is not about creating a collapse; rather, it means opening up opportunities for intentional change and making them significantly more visible. Based on these criteria, I define the following as triggers: the 2018 IPCC report, the introduction of the GND (2018–2019),[9] climate activism (2018–2019), BLM protests (2020), and the COVID-19 pandemic, including the economic recession following it (2020). In one form or another, all of these elements constitute either an exogenous shock or a major deliberate social disruption (see Table 5.1).

Figure 5.1 presents the triggers, the connections between them, and the connections with other abovementioned trends and events. Some of these connections are stronger than others (e.g., the one between climate activism and the GND, or between COVID-19 and the economic downturn following it and the green recovery plans).

It should be noted that ecological disruption could also be a trigger in the case of "reaching a tipping point and moving toward a new regime that is entirely undesirable by anyone within the social system" (Moore et al. 2014, 3). However, I do not consider any of the recent escalating climate change impacts to be an ecological disruption. Even though the Earth system has become more fragile (Carpenter et al. 2019) and we seem to be moving toward a catastrophic ecological tipping point (Lenton et al. 2019), we have not yet reached it.

Three Potential Pathways for Transformation

The next phase in the transformation journey, which serves as a preparation stage, includes three processes: sensemaking, envisioning, and gathering momentum. Moore et al.'s (2014) understanding of sensemaking is based on Gioia's (1986) definition as a process in which

---

[9] It also includes the introduction of the European Green Deal in 2019 (European Commission 2019), which has many similarities with the GND. While the European Green Deal may not be as socially ambitious as its American counterpart, it is committed, for example, to ensure that the transition to a net-zero economy will be made in a just and fair manner.

Table 5.1   The five triggers explained

Elements (Event/trend)[10]	Criteria 1: Exogenous shock	Criteria 2: A deliberate social disruption
2018 IPCC report	No	Yes. It changed the perceptions about the need for a 1.5 °C limit and the urgency to act, offering a science-based roadmap to limit warming to 1.5 °C
Introduction of Green New Deal (and the European Green Deal; 2018–2019)	No	Yes. It went against the notion that a climate policy should focus only on decarbonizing the economy by offering a holistic climate policy agenda that integrates just transition and other social and economic components with climate goals. It makes the case that only a big plan can be sufficient to address the climate challenge
Climate activism (2018–2019)	No	Yes. It helped define climate change as a crisis. It is exposing the insufficiency of the current response to this crisis while empowering people (young people in particular) to reject inadequate institutional and political responses to it
Black Lives Matter protests (2020)	No	Yes. It shed a light on the prominence of inequality and injustice in the system, pushing racial justice from the periphery to the core of the climate agenda
COVID-19 and the economic downturn (2020)	Yes	COVID-19 has disruptive features, but it is not an example of a deliberate social disruption. The pandemic and the economic downturn also created significant opportunities for intentional change (e.g., green recovery plans)

individuals "construct meaningful explanations for situations and their experiences within those situations" (61). According to Moore et al. (2014), envisioning means developing alternative pathways and new innovations/visions for a preferable future, while gathering momentum entails mobilization around the new ideas envisioned and experimentation with them in "protected" niches. Looking at the different manifestations of these processes which have evolved over the last few years,[11] I see three

---

[10]While the GND and climate activism continued to play a role in 2020, they were functioning as a trigger mainly in 2018–2019. Therefore, this is the timeframe used for these elements in the table.

[11] The time frame of phase 2 is similar to that of phase 1, which is aligned with Moore et al.'s (2014) suggestion that "the phases may occur simultaneously or in varying order in any transformation process" (3).

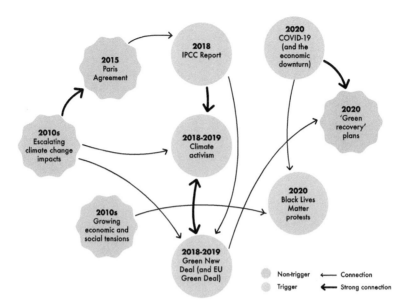

Fig. 5.1 Connections between the triggers and other previously identified trends/events

potential pathways for the transformation process: fixing, reforming, and revolutionizing the system (see Table 5.2).

The first pathway focuses almost entirely on decarbonizing the economy and accelerating its transition into a zero-carbon one. It is aligned with and driven by the need to limit warming to 1.5 °C and is based on the belief that providing companies with clear roadmaps, pathways, and playbooks on how to move forward in this journey will motivate more of them to get on board. This approach also champions a virtuous cycle, whereby companies are mobilized to set (voluntary) ambitious decarbonization goals and influence governments to set more ambitious policies, which in return will push companies further to deliver bold action. Overall, the focus is on developing strategies, which consider climate change as a technical problem that "can be successfully addressed by applying greater expertise, more innovation, and better management" (O'Brien 2018, 154). Examples for this approach include the We Mean

Table 5.2 Phase 2's three processes

Process/Pathway	1. Fixing the system	2. Reforming the system	3. Revolutionizing the system
Sensemaking	Climate change is seen as a technical problem, which requires rapid decarbonization of the economy to limit warming to a maximum of 1.5 °C	The scale of climate change dictates a forceful government intervention that will drive wide-scale social mobilization, addressing both climate change and social inequality, as these issues cannot be separated	Capitalism is at the root of climate change; therefore, to solve the problem it needs to be rejected and replaced with a better system
Envisioning	The business sector sets ambitious targets and takes bold action to meet the 1.5 °C goal. It also drives more ambitious government action on climate to accelerate the shift to a zero-carbon economy	A comprehensive, government-led upgrade and reform of every part of the economy is undertaken, including a just and fair transition, to fight climate change and systemic injustices at once	A range of post-capitalistic visions is developed, articulating alternative practices (e.g., alternative production/finance, platform coops) and routes to a post-capitalistic society (e.g., solidarity economy, doughnut economics)
Gathering momentum	Growing activity of coalitions/networks of companies that are committed to bold climate action in line with the 1.5 °C ambition and that work with other actors to accelerate the transition to a zero-carbon economy. Examples include the We Mean Business coalition, the Science Based Targets initiative (SBTi), and the Exponential Roadmap Initiative	Political mobilization occurs around the policy agenda of the GND and the European Green Deal. Experiments with GND-aligned policy tools are already taking place at state and local levels in the U.S. (Mccoy 2019)	Ideas challenging capitalism, particularly degrowth (e.g., Cassidy 2020; Roulet and Bothello 2020), have been moving from the periphery to the mainstream, with growing support by climate activists (e.g., Extinction Rebellion), academics and members of the European Parliament. There are also experiments with the doughnut economics framework on a city level (Raworth 2020)[12]

[12] In 2020, Amsterdam became the first city to embrace the doughnut economics framework—see https://rb.gy/n8py1a.

Business Coalition,[13] the Science Based Targets initiative (SBTi),[14] and the Exponential Roadmap Initiative.[15]

The second pathway acknowledges that the climate crisis cannot be addressed separately from other social and economic challenges (such as inequality and systemic injustices), as they are all connected. It is interested in government-led, comprehensive solutions which will transform the economy "so that it is less resource-intensive, and … do it in ways that are equitable, with the most vulnerable protected and the most responsible bearing the bulk of the burden" (Klein 2020, 33). The best examples of this approach are the GND in the U.S. (Ocasio-Cortez 2019) and the European Green Deal (European Commission 2019).[16] As this approach focuses on policy changes, the efforts to gather momentum around it are mainly political, including efforts by climate activists such as the Sunrise Movement to push the GND further into the political mainstream.

While the GND aims to reduce emissions "through economic transformation" (Ocasio-Cortez 2019, 3), for some it does not go far enough in terms of challenging the economic system. This notion is at the core of the third pathway, which believes that addressing climate change requires a more radical approach that would truly challenge the fundamental principles of capitalism, which it considers to be the root cause of the climate crisis (Simon 2019). As Hickel (2020) explains: "our addiction to fossil fuels, and the antics of the fossil fuel industry, is really just a symptom of a prior problem. What's ultimately at stake is the economic system that has come to dominate more or less the entire planet over the past few centuries: capitalism" (19).

The third pathway is grounded in concepts such as *Capitalocene*, "which signifies capitalism as a way of organizing nature—as a multi-species, situated, capitalist world-ecology" (J. W. Moore 2016, 6)[17]

---

[13] See https://www.wemeanbusinesscoalition.org/.

[14] See http://sciencebasedtargets.org/.

[15] See https://exponentialroadmap.org/.

[16] It may seem somewhat unusual that the GND and the European New Deal are shown as examples of both phase 1 and phase 2, but nevertheless they embody characteristics of both phases.

[17] Moore (2016) juxtaposes Capitalocene with Anthropocene, "a new conceptualization of geological time—one that includes 'mankind' as a 'major geological force'" (3), which he believes to be important yet inadequate as it ignores the critical impact of capitalism on the Earth's ecosystems.

and *degrowth*, "a planned downscaling of energy and resource use to bring the economy back into balance with the living world in a safe, just and equitable way" (Hickel 2020, 29). The latter is also one of the key points of disagreement with the GND, which seems to be more aligned with the "green growth" that proponents of degrowth find antithetical to a 1.5 °C ambition (Mastini 2020; Hickel 2020). Offering different visions of a post-capitalistic economy (e.g., Pirani 2020; Alexander 2020), the third pathway is supported by scholars as well as by political and climate movements (e.g., the Democratic Socialists of America,[18] ExtinctionRebellion).

When Vulnerabilities and Possibilities Become Clearer

As of now, we do not seem to have moved beyond the second phase in the transformation process: we are still in the preparation stage, where we experiment with different ideas, innovations, and visions. Overall, the aforementioned pathways that compete for dominance in this process (fixing, reforming, and revolutionizing) are still in the early stages of experimentation for the most part, with some potential for synergies; for example, between the GND and degrowth (Mastini et al. 2021).

While being in a preparation phase only may seem somewhat disappointing to those hoping for a quicker transformation, we need to acknowledge the significant advancement in the process made in the last few years. Even if 2020 had not been as significant as it turned to be (with COVID-19, global recession, BLM protests, and so on), I do believe that the transformation process would still have materialized in this decade, given the combination of growing pressure to act on climate change and the accelerated climate impacts. Yet it seems plausible that the resistance to major changes in the system would push toward a "lighter" version of transformation. In other words, this transformation would mostly be about fixing the system in relation to climate change, rather than reforming or revolutionizing it. Furthermore, it would probably be a slower process to allow companies a more comfortable transition.

The year 2020 seems to be changing this expected trajectory. While it may be too soon to evaluate its full impacts, 2020 can both accelerate the transformation process and make it more progressive (i.e., we may

---

[18] See https://www.dsausa.org/.

have a transformation that is more grounded in the second or third pathways). This trajectory change could be influenced by the vulnerabilities and possibilities which this year illuminated.

Events in 2020, in particular the advent of COVID-19, exposed many vulnerabilities in the current economic model, including the growing exposure of value chains to shocks (McKinsey Global Institute 2020), the widening worldwide gap between rich and poor (Cugat and Narita 2020; BBC News 2020), and inequality disparities along racial and ethnic lines (Buckley and Barua 2020). In contrast were the many examples of making the impossible possible. We saw things on a global scale that were almost unimaginable until then, such as the suspension of most air travel, shut-downs of much of the economy in many countries, and blue skies and fresh air in highly polluted cities. As Rebecca Solnit (2020) observed: "things that were supposed to be unstoppable stopped, and things that were supposed to be impossible—extending workers' rights and benefits, freeing prisoners, moving a few trillion dollars around in the US—have already happened" (para. 1).

The vulnerabilities and possibilities generated by this "perfect storm" of crises prompted even McKinsey Global Institute to publish a paper questioning the future of capitalism, no less (Manyika, Pinkus, and Tuin 2020). The authors suggested that COVID-19 "creates an opportunity to reconsider how well the current model works and the extent to which it is fit for the challenges and opportunities of the present and future" (9). While these questions are not necessarily novel, the context in which they are asked is, and the vividness of both the vulnerabilities of our systems to shocks and the possibilities of doing things differently may increase the political and societal acceptance of ideas that until lately were considered unacceptable. The consequences may be greater openness and willingness to pursue a transformational change that pushes the envelope much further than just fixing the system. This sentiment is consistent with O'Brien's (2020) assertion that this is "a significant moment and we now have the opportunity to break with the momentum of the past and engage with new ways of thinking and being" (5).

Overall, while there is no clear timeline for the next phases of the transformation journey, the events of 2020 seem to make it more plausible that this decade may open the way for a more far-reaching and faster process of transformation. For companies, it may be useful to consider this moment as a strategic inflection point, which Grove (1996) describes as "a time in the life of a business when its fundamentals are about to change. That change can mean an opportunity to rise to new heights. But it may just as likely signal the beginning of the end" (3). Rita McGrath (2019) states that this point "dramatically shifts some element of your activities, throwing certain taken-for-granted assumptions into question" (1). This description seems to be a good fit for this moment and what is about to come for companies, especially with regard to corporate sustainability.

References

Alexander, Samuel. 2020. "Post-Capitalism by Design Not Disaster." *The Ecological Citizen* 3(Suppl B): 13–21. https://rb.gy/niyztx.

Arias-Maldonado, Manuel. 2020. "Sustainability in the Anthropocene: Between Extinction and Populism." *Sustainability* 12 (6): 2538. https://doi.org/10.3390/su12062538.

BBC News. 2020. "Coronavirus: BBC Poll Suggests Stark Divide Between Rich and Poor Countries." *BBC News*, September 10. https://www.bbc.com/news/world-54106474.

Beeson, Mark. 2019. "Can Environmental Populism Save the Planet?" *The Conversation*, August 20. https://rb.gy/lojjfg.

Blankenship, Mary, and V. Richard Reeves. 2020. "From the George Floyd Moment to a Black Lives Matter Movement, in Tweets." *The Brookings Institution*, July 10. https://rb.gy/wpgg0d.

Buchanan, Larry, Quoctrung Bui, and Jugal K. Patel. 2020. "Black Lives Matter May be the Largest Movement in U.S." *The New York Times*, July 3. https://rb.gy/s20llg.

Buckley, Patricia, and Akrur Barua. 2020. "COVID-19's Impact on Income Inequality." *Deloitte Insights*, July. https://rb.gy/7ixdbg.

Carpenter, Stephen R., Carl Folke, Marten Scheffer, and Frances R. Westley. 2019. "Dancing on the Volcano: Social Exploration in Times of Discontent." *Ecology and Society* 24 (1). https://doi.org/10.5751/ES-10839-240123.

Cassidy, John. 2020. "Can We Have Prosperity Without Growth?" *The New Yorker*, February 3. https://rb.gy/8s5zdr.

Chen, Wenjie, Mico Mrkaic, and Malhar Nabar. 2019. "The Global Economic Recovery 10 Years After the Crisis." IMF Working Paper 19/83. Washington, DC. https://rb.gy/1itzds.

Collins, Chuck, Dedrick Asante-Muhammed, Josh Hoxie, and Sabrina Terry. 2019. "Dreams Deferred: How Enriching the 1% Widens the Racial Wealth Divide." Washington, DC. https://rb.gy/gxvus0.

Cugat, Gabriela, and Futoshi Narita. 2020. "How COVID-19 Will Increase Inequality in Emerging Markets and Developing Economies." *IMF Blog*, October 29. https://rb.gy/lubhdk.

Dyer-Witheford, Nick. 2020. "Left Populism and Platform Capitalism." *Triplec* 18 (1): 116–31. https://doi.org/10.31269/triplec.v18i1.1130.

European Commission. 2019. "A European Green Deal." European Commission. https://ec.europa.eu/info/strategy/priorities-2019-2024/eur opean-green-deal_en.

Figueres, Christiana, and Tom Rivett-Carnac. 2020. *The Future We Choose: Surviving the Climate Crisis*. New York: Alfred A. Knopf.

Folke, Carl, Reinette Biggs, Albert V. Norström, Belinda Reyers, and Johan Rockström. 2016. "Social-Ecological Resilience and Biosphere-Based Sustainability Science." *Ecology and Society* 21 (3). https://doi.org/10.5751/ES-08748-210341.

Freedman, Andrew, Jason Samenow, Rick Noack, and Karly Domb Sadof. 2020. "Climate Change in the 2010s: Decade of Fires, Floods and Scorching Heat Waves—The Washington Post." *The Washington Post*, January 1. https://rb. gy/atgwxz.

Gardiner, Beth. 2019. "For Europe's Far-Right Parties, Climate is a New Battleground." *Yale E360*, October 29. https://e360.yale.edu/features/for-eur opes-far-right-parties-climate-is-a-new-battleground.

Gioia, D. A. 1986. "Symbols, Scripts, and Sensemaking: Creating Meaning in the Organizational Experience." In *The Thinking Organization*, edited by H. Sims D. A. Gioia, 49–74. San Francisco, CA: Jossey-Bass.

Grove, Andrew S. 1996. *Only the Paranoid Survive: How to Exploit the Crisis Points That Challenge Every Company*. New York: Doubleday.

Hall, C Michael, Daniel Scott, and Stefan Gössling. 2020. "Pandemics, Transformations and Tourism: Be Careful What you Wish For." *Tourism Geographies* 22 (3): 577–98. https://doi.org/10.1080/14616688.2020.1759131.

Heeb, Gina. 2019. "US Income Inequality Jumps to Highest Level Ever Recorded." *Business Insider*, September 27. https://rb.gy/wxm4di.

Hickel, Jason. 2020. *Less is More: How Degrowth Will Save the World*. London: William Heinemann.

ILO, FAO, IFAD, and WHO. 2020. "Impact of COVID-19 on People's Livelihoods, Their Health and our Food Systems." *World Health Organization*, October 13. https://rb.gy/t9liun.

IPCC. 2012. "Managing the Risks of Extreme Events and Disasters to Advance Climate Change Adaptation—Summary for Policymakers." In *A Special Report of Working Group I and II of the Intergovernmental Panel on Climate Change*. https://rb.gy/pnr2rk.

———. 2018a. "Global Warming of 1.5 °C." https://www.ipcc.ch/sr15/.

———. 2018b. "Summary for Policymakers of IPCC Special Report on Global Warming of 1.5 °C Approved by Governments." *IPCC*, October 8. https:// rb.gy/fxsuzh.

Irfan, Umair. 2020. "Disasters 2020: Fires, Floods, Hurricanes, Typhoons, and Locusts Set Records." *Vox*, December 30. https://rb.gy/zspeza.

Klein, Naomi. 2020. "Market Fundementalism at the Worst Time." In *Winning the Green New Deal: Why We Must, How We Can*, edited by Varshini Prakash and Guido Girgenti. New York: Simon and Schuster.

Knuth, Katherine. 2019. *How Collectives Drive Deliberate Transformation to Make Progress Toward Sustainability*. University of Minnesota.

Leichenko, Robin, and Karen O'Brien. 2019. *Climate and Society: Transforming the Future*. Cambridge, UK: Polity Press.

Lenton, Timothy M., Johan Rockström, Owen Gaffney, Stefan Rahmstorf, Katherine Richardson, Will Steffen, and Hans Joachim Schellnhuber. 2019. "Climate Tipping Points—Too Risky to Bet Against." *Nature*. Nature Research. https://doi.org/10.1038/d41586-019-03595-0.

Lockwood, Matthew. 2018. "Right-Wing Populism and the Climate Change Agenda: Exploring the Linkages." *Environmental Politics* 27 (4): 712–32. https://doi.org/10.1080/09644016.2018.1458411.

Manyika, James, Gary Pinkus, and Monique Tuin. 2020. *Rethinking the Future of American Capitalism | McKinsey*. McKinsey Global Institute, November 12. https://rb.gy/wu6f3k.

Mastini, Riccardo. 2020. "A Post-Growth Green New Deal." *Uneven Earth*, February 17. http://unevenearth.org/2020/02/a-post-growth-green-new-deal/.

Mastini, Riccardo, Giorgos Kallis, and Jason Hickel. 2021. "A Green New Deal Without Growth?" *Ecological Economics* 179 (January): 106832. https://doi.org/10.1016/j.ecolecon.2020.106832.

Mccoy, Caitlin. 2019. *The States as Green new Deal Policy Labs*. Cambridge, MA. https://rb.gy/trfinm.

McGrath, Rita Gunther. 2019. *Seeing Around Corners: How to Spot Inflection Points in Business Before They Happen*. New York: Houghton Mifflin Harcourt.

McKinsey & Company. 2020. *COVID-19: Implications for Business*. McKinsey & Company, October 21. https://rb.gy/fx64ri.

McKinsey Global Institute. 2020. "Risk, Resilience, and Rebalancing in Global Value Chains." https://rb.gy/x3pidj.

Moore, Jason W. 2016. "Introduction: Anthropocene or Capitalocene? Nature, History, and the Crisis of Anthropocene or Capitalocene? Nature, History, and the Crisis of Capitalism Capitalism." In *Anthropocene or Capitalocene? Nature, History, and the Crisis of Anthropocene or Capitalocene? Nature, History, and the Crisis of Capitalism Capitalism*, edited by Jason W. Moore. Oakland, CA: PM Press.

Moore, Michele Lee, Ola Tjornbo, Elin Enfors, Corrie Knapp, Jennifer Hodbod, Jacopo A. Baggio, Albert Norström, Per Olsson, and Duan Biggs. 2014.

"Studying the Complexity of Change: Toward an Analytical Framework for Understanding Deliberate Social-Ecological Transformations." *Ecology and Society* 19 (4). https://doi.org/10.5751/ES-06966-190454.

Nalau, Johanna, and John Handmer. 2015. "When is Transformation a Viable Policy Alternative?" *Environmental Science and Policy* 54: 349–56. https://doi.org/10.1016/j.envsci.2015.07.022.

O'Brien, Karen. 2012. "Global Environmental Change II: From Adaptation to Deliberate Transformation." *Progress in Human Geography* 36 (5): 667–76. https://doi.org/10.1177/0309132511425767.

————. 2018. "Is the 1.5 °C Target Possible? Exploring the Three Spheres of Transformation." *Current Opinion in Environmental Sustainability* 31 (April): 153–60.

————. 2020. "You Matter More Than You Think: Quantum Social Change in Response to a World in Crisis." Unpublished manuscript circulated for feedback in June 2020. Oslo, Norway.

O'Brien, Karen, and Linda Sygna. 2013. "Responding to Climate Change: The Three Spheres of Transformation." In *Proceedings of Transformation in Changing Climate International Conference*, 16–23. Oslo, Norway: University of Oslo.

Ocasio-Cortez, Alexandria. 2019. *H.Res.109—Recognizing the Duty of the Federal Government to Create a Green New Deal*. Washington DC: 116th Congress (2019–2020). https://www.congress.gov/bill/116th-congress/house-resolution/109/text.

Olsson, Per, Carl Folke, and Thomas Hahn. 2004. "Social-Ecological Transformation for Ecosystem Management: The Development of Adaptive Co-Management of a Wetland Landscape in Southern Sweden." *Ecology and Society* 9 (4).

Park, S. E., N. A. Marshall, E. Jakku, A. M. Dowd, S. M. Howden, E. Mendham, and A. Fleming. 2012. "Informing Adaptation Responses to Climate Change through Theories of Transformation." *Global Environmental Change* 22 (1): 115–26. https://doi.org/https://doi.org/10.1016/j.gloenvcha.2011.10.003.

Partington, Richard. 2019. "Inequality: Is It Rising, and Can We Reverse It?" *The Guardian*, September 9. https://rb.gy/zynvx0.

Pastor, Lubos, and Pietro Veronesi. 2020. "Inequality Aversion, Populism, and the Backlash Against Globalization." Chicago Booth Research Paper No. 20–11. https://doi.org/10.2139/ssrn.3224232.

Pirani, Simon. 2020. "Socialism, Capitalism and the Transition Away From Fossil Fuels." *OpenDemocracy*, January 28. https://rb.gy/ydbzke.

Raworth, Kate. 2020. "So You Want to Downscale the Doughnut? Here's How." *Kate Raworth*, July 16. https://rb.gy/nqh21r.

Roulet, Thomas, and Joel Bothello. 2020. "Why 'De-Growth' Shouldn't Scare Businesses." *Harvard Business Review*, February 14. https://hbr.org/2020/02/why-de-growth-shouldnt-scare-businesses.

Sedik, Tahsin Saadi, and Rui Xu. 2020. "A Vicious Cycle: How Pandemics Lead to Economic Despair and Social Unrest." IMF Working Paper 20/216. Washington, DC. https://rb.gy/x1rjvv.

Simon, Matt. 2019. "Enter the Capitalocene: How Climate Change Will Ruin Capitalism." *WIRED*, September 20. https://www.wired.com/story/capitalocene/.

Solnit, Rebecca. 2020. "'The Impossible Has Already Happened': What Coronavirus Can Teach Us about Hope." *The Guardian*, April 7. https://rb.gy/kxq1sk.

The Club of Rome. 2020. "Planetary Emergency 2.0 Securing a New Deal for People, Nature and Climate." Winterhur, Switzerland. https://rb.gy/l37v1a.

Thunberg, Greta. 2019. "'Our House Is On Fire': Greta Thunberg, 16, Urges Leaders to Act on Climate." *The Guardian*. https://rb.gy/i86mwo.

Triggs, Adam, and Homi Kharas. 2020. "The Triple Economic Shock of COVID-19 and Priorities for an Emergency G-20 Leaders Meeting." Brookings. 2020.

UN DISD. 2020. "The Social Impact of COVID-19." United Nations. 2020.

UNEP. 2020. "Emissions Gap Report 2020." Nairobi. https://rb.gy/ehhfom.

UNFCCC. 2015. "The Paris Agreement." *United Nations Climate Change*. https://unfccc.int/process-and-meetings/the-paris-agreement/the-paris-agreement.

Verbeke, Alain. 2020. "Will the COVID-19 Pandemic Really Change the Governance of Global Value Chains?" *British Journal of Management* 31 (3): 444–46. https://doi.org/10.1111/1467-8551.12422.

Weitzman, Hal. 2020. "The Populism Puzzle." *Chicago Booth Review*, March 2. https://review.chicagobooth.edu/economics/2020/article/populism-puzzle.

WMO. 2020. "WMO Statement on the State of the Global Climate in 2019." WMO-No. 1248. Geneva, Switzerland. https://library.wmo.int/doc_num.php?explnum_id=10211.

Worland, Justin. 2020. "America's Long Overdue Awakening on Systemic Racism." *Time*, June 11. https://time.com/5851855/systemic-racism-america/.

CHAPTER 6

The Vision: Awakened Sustainability

Abstract This chapter lays out a vision for the mode that will replace sustainability-as-usual. Called "awakened sustainability," this vision is driven by a societal awakening to the reality of climate crisis and social injustice. The chapter outlines in detail a number of elements critical for the creation of a vision (mental model, vision content, and vision context), as well as the backcasting process, of which the vision is a part. The vision of awakened sustainability is designed to give new life to the concept of sustainability, in consideration of the urgency of the climate crisis as well as the need to prioritize social justice and apply regenerative design in the formation of sustainable solutions. It sets ambitious principles and is grounded in a mental model ("Sustainability first, NOW") that is juxtaposed against sustainability-as-usual using the "opposite principle": if we consider every instinct ingrained in sustainability-as-usual to be wrong, then the opposite would indeed be right.

Keywords Social justice · Regenerative design · Backcasting · Awakened sustainability · Vision · Sustainable design

"We cannot achieve what we cannot imagine"—*Elise Boulding.* (Boulding 2017, 120)

101

R. Godelnik, *Rethinking Corporate Sustainability in the Era of Climate Crisis*, https://doi.org/10.1007/978-3-030-77318-2_6

In the last chapter, I discussed a transformation process that is now unfolding, driven by a convergence of social and economic pressures, an exogenous shock (i.e., COVID-19), and climate change. I consider this process, or journey if you will, to be a driving force that can reshape corporate sustainability by helping moving companies from their current state of sustainability-as-usual to a new desired state I define as *awakened sustainability*.

This chapter aims to lay out a vision for this new state of awakened sustainability to help bridge between the future and the present. My focus is on outlining a clear vision for a desired future to gain a better understanding of where companies should be heading throughout their sustainability journey, both over this decade and beyond. Outlined below are the elements that are critical for the formation of a vision (mental model, vision content and context), as well as the backcasting process, of which the vision is a part.

BACKCASTING

The starting point requires an articulation of the backcasting process I will be using, as it will help illuminate why we need a vision in the first place. *Backcasting* can be defined as "generating a desirable future, and then looking backwards from that future to the present in order to strategize and to plan how it could be achieved" (Vergragt and Quist 2011, 747). Similarly to forecasting, it is used to support planning and decision-making processes. However, forecasting extrapolates current trends and uses historical data to develop likely future scenarios, whereas backcasting is concerned "not with what futures are likely to happen, but with how desirable futures can be attained" (Robinson 1990, 822).

Dreborg (1996) argues that forecasting "is unlikely to generate solutions that would presuppose the breaking of trends" (814) because it is based on dominant trends. He adds that backcasting is especially useful with complex problems, when a major change is needed, when the dominant trends are part of the problem, when externalities are key to the problem, and when "the time horizon is long enough to allow considerable scope *for deliberate choice*" (816). Sustainability challenges meet these criteria, which suggests that backcasting could be useful in addressing them (Robèrt 2002). Vergragt and Quist (2011) note that the normative nature of both sustainability and backcasting (which considers the desired future) also supports the case for applying backcasting to sustainability issues; thus, it is not surprising to see a growing interest in doing so (Quist 2007).

Out of the different backcasting approaches (e.g., see Quist 2007), I choose to build on The Natural Step's approach, which was developed in collaboration with a network of scientists and business partners (Holmberg 1998). Called the "ABCD procedure" (or process), it is part of a broader "Framework for Strategic Sustainable Development" (FSSD), also developed by The Natural Step (Broman and Robèrt 2017). The ABCD process includes four steps: (A) developing a vision of success for sustainability (framed by sustainability principles); (B) assessing the current situation vis-à-vis the vision; (C) ideating possible steps and solutions; and (D) prioritizing steps and choosing a path forward. This chapter considers step A, while the following chapters loosely follow the other steps.

DEVELOPING A VISION

Figueres and Rivett-Carnac (2020) write that "a compelling vision is like a hook in the future. It connects you to the pockets of possibility that are emerging and helps you pull them into the present" (99). Furthermore, a compelling vision could be the starting point for a transformational journey toward a sustainable future. As Meadows et al. (1992) note, "vision without action is useless. But action without vision does not know where to go or why to go there. Vision is absolutely necessary to guide and motivate action" (224). Yet as necessary as it may seem, what is this vision all about?

To discern that, we first need to consider what constitutes a *vision*. Although the term has many definitions, I find those related to organizational contexts most useful in the context of this book. First, Senge's (1990) point about a shared vision being "a picture of the future we seek to create" (9) provides a good understanding of vision on the most basic level. Strange and Mumford's (2005) suggestion that a vision "involves a set of beliefs about how people should act, and interact, to make manifest some idealized future state" (122) also helps to characterize its nature. They also connect vision to mental models by suggesting that a prescriptive model provides the foundation for a vision because it reflects the system as it should be, whereas normative or descriptive ones describe the system as it currently is (Mumford and Strange 2002). This notion goes along with our general understanding of the importance of mental models, which "determine not only how we make sense of the world, but how we take action" (Senge 1990, 164). Therefore, defining a mental model constitutes an important step in the process of forming a vision.

A New Mental Model: Sustainability first, NOW

Thinking about a vision in terms of "a picture of the future we seek to create" (Senge 1990, 9), we can consider mental models as the lenses we put on to create this picture. A useful metaphor for explaining the lenses for awakened sustainability comes from the popular 1990s TV show "*Seinfeld*." In a 1994 episode entitled "The Opposite," Jerry Seinfeld and George Costanza have an "aha!" moment at the coffee shop:

> *George: It all became very clear to me sitting out there today that every decision I have ever made in my entire life has been wrong. My life is the complete opposite of everything I wanted it to be. Every instinct I have, in every aspect of life, be it something to wear, something to eat... It's all been wrong.*
>
> ....
>
> *Jerry: If every instinct you have is wrong, then the opposite would have to be right.*
>
> *George: Yes, I will do the opposite. I used to sit here and do nothing, and regret it for the rest of the day, so now I will do the opposite, and I will do something!* (Seinfeld et al. 1994)

To some degree, the lenses I use to form the vision are based on the same "opposite principle": if we consider every instinct ingrained in sustainability-as-usual to be wrong, then the opposite would indeed be right. In this case, it means that companies need to flip their priorities and consider sustainability first and profits last, not the other way around (as done today). As explained in detail in Chapter 2, sustainability-as-usual is grounded in a mindset defined as shareholder capitalism 2.0, whereby companies' economic considerations of profit maximization transcend those of sustainability. This mental model has been echoed in the quest to make the business case for sustainability. Here companies try to figure out how pursuing sustainable strategies and practices can benefit their bottom line, under the assumption that profit maximization tops all other considerations. Defined as "sustainability first, NOW," my "opposite" mental model reverses these dynamics, assuming that sustainability considerations top everything else (see Fig. 6.1). As a result, rather than

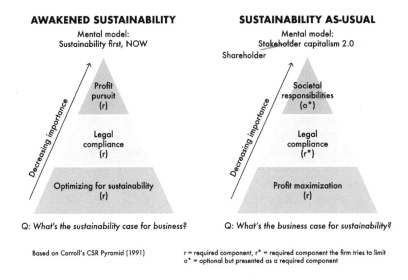

Fig. 6.1 Sustainability-as-usual vs. awakened sustainability

making the business case of sustainability, companies will start making the sustainability case of business.[1]

This new mental model has some resemblance to Fletcher and Tham's (2019) suggestion to "place earth first, before profit" (14).[2] While sustainability will be prioritized, the pursuit of profits will remain a required condition for companies to succeed in the new state of awakened sustainability, although it will no longer be the key element for which they are optimized. Companies will experience a shift from their current state, wherein they may be doing good to do well but "continue to do bad when that is the best way for them to do well" (Bakan 2020, 27). In their new state, not only is doing bad no longer an option, but sustainability is no longer a means to achieve profits but the end goal. If and when conflicts arise between sustainability and profits, companies

---

[1] This approach echoes Ergene et al.'s (2020) call to pursue a more radical research agenda "that takes us beyond traditional theories, models, and frameworks," including a shift from the business case for sustainability to the ecological case for business.

[2] Fletcher and Tham's suggested systemic change (in the fashion sector) also addresses economic growth, making the case for growing out of growth and replacing growth logic with earth logic. On the other hand, I offer a change that is more growth agnostic.

must be creative and innovative in order to align themselves with the new power dynamics.

In theory, the new mindset could be agnostic to growth. To paraphrase Kate Raworth (2017), I mean agnostic in the sense of designing companies that promote sustainability whether their profits increase, decrease, or remain steady.[3] As such, the new mental model can work with the application of different strategies as long as they adhere to the shift in priorities, so that sustainability always comes first.

The new mental model is grounded not only in a growing understanding of the shortcomings of the current model of sustainability-as-usual, but also in the urgent need to act on climate change, which is why it includes the word "NOW" in capitals. Already well established and supported by the latest IPCC reports (2018, 2019), the urgency of addressing climate change is becoming a defining characteristic of our time, reflecting the perspective that "winning slowly is the same as losing" (McKibben 2017). Not only are we in a fairly dire state at present, but "climate change is running faster than we are," as the UN Secretary-General António Guterres (2018, para. 3) noted. In these circumstances, winning in terms of climate change requires moving with greater speed while adopting a completely different approach (i.e., the opposite of the status quo). Moreover, we should consider that the level of warming already locked into the system will inevitably cause more problematic changes in our climate (Buis 2019); therefore, we will likely need to act swiftly for a very long time, not just for a limited transition period.

Ultimately, the new mindset of "sustainability first, NOW" aims to allow for new possibilities. As Wright claims, "the actual limits of what is achievable depend in part on the beliefs people hold about what sorts of alternatives are viable" (Wright 2010, 15). Thus, adopting an opposite approach to corporate sustainability and grounding it in a sense of urgency should help release companies from the chains of the current mental model, which results in very narrow "limits of possibility," as Wright (2010) observed.

---

[3] Raworth refers to being agnostic to growth in GDP. Her exact quote is as follows: "By agnostic, I do not mean simply not caring whether GDP growth is coming or not, nor do I mean refusing to measure whether it is happening or not. I mean agnostic in the sense of designing an economy that promotes human prosperity whether GDP is going up, down, or holding steady" (Raworth 2017, 209).

Vision's Content and Context

The mental model, as stated earlier, is just the foundation for the vision. So while using these lenses helps one imagine the picture of a desired future more clearly, the vision (i.e., the picture) is still incomplete without concise content and context to define it. The content provides clear boundaries for corporate sustainability, and the context sheds light on the issues and approaches to be prioritized. Continuing to build on Senge's (1990) metaphor of a vision as a picture, we can consider the context as the paintbrush we use to paint the picture.

(1) Principle-based content

The approaches that could be applied to the challenge of forming a vision are various (e.g., see Lockton and Candy 2018). For example, a compelling vision for a sustainable future could be made using images, stories, experiences, scenarios, and even policy proposals (e.g., the Green New Deal[4]). Given the growing level of change and uncertainty, Broman and Robèrt (2017) suggest that it is better to adopt a more flexible and open-ended form of vision, such as one based on principles. They make the case that "backcasting from visions framed by a principled definition of sustainability is a more generic, intuitive, and practical approach for supporting sustainable development" (Broman and Robèrt 2017, 20).

Aligned with their approach, I would offer a principle-based vision. While it may not seem as rich in detail as a scenario-based one, for example, it provides a greater level of flexibility and more opportunities to evolve as needed. These aspects seem valuable, especially for a vision focusing on sustainability. A principle-based vision can be seen as the scaffolding to support the construction of various sustainable structures in a corporate setting, as long as they align with the predefined principles.

Once we establish that the content is based on principles, the question follows as to what these should be exactly. The search starts with the work conducted by The Natural Step on developing the FSSD to define sustainability rigorously and provide guidance on how to realize a vision grounded in that definition (Broman and Robèrt 2017). The Natural

---

[4] The 2019 resolution on the Green New Deal (Ocasio-Cortez 2019) was followed by a short film narrated by Congresswoman Alexandria Ocasio-Cortez, which showed what its successful implementation could look like (see https://youtu.be/d9uTH0iprVQ).

Step's quest in this regard yielded four principles, which "could be used as a non-perspective starting point for system thinking about sustainability" (Holmberg 1998, 31). Also referred to as "system conditions," they are based on a consensus of the researchers involved in the work on the FSSD as to what exactly is needed to achieve sustainable development. The principles suggest that for a society to be sustainable, nature should not be subjected to systematic increase in: "(1) concentrations of substances from the Earth's crust; (2) concentrations of substances produced by society; (3) degradation by physical means, and, in that society ... (4) people are not subject to conditions that systematically undermine their capacity to meet their needs" (Ny et al. 2008, 64). The last principle, which considers the social dimension of sustainability, was later developed into five distinct principles, which convey that people in a sustainable society should not be subject to structural obstacles to (1) health, (2) influence, (3) competence, (4) impartiality, and (5) meaning-making (Missimer et al. 2010).

Next, I look into the use of the FSSD sustainability principles by the Future-Fit Foundation, which considers them "the 'rules of the game' to which we must all adhere" (Future-Fit Foundation 2020, 23). According to the Future-Fit Foundation, these principles "indicate what patterns of behaviour are environmentally and socially acceptable, in the sense that they avoid causing degenerative outcomes" (23). With a vision of transforming companies (and eventually society) to become future-fit or truly sustainable, the Future-Fit Foundation has drawn on the FSSD sustainability principles to develop properties of a future-fit society. Essentially, it "translates" them into more actionable qualities (or properties), which will ensure "the possibility that humans and other life will flourish on Earth forever, by being environmentally restorative, socially just, and economically inclusive" (Kendall and Rich 2019, 244).

The Future-Fit Foundation considers the exclusion criteria of the FSSD sustainability principles as a strength: "this should be seen as liberating rather than restrictive: as long as they are not breached, anything is possible" (Future-Fit Foundation 2020, 24). Nevertheless, it has transformed them into properties framed in positive terms, as "regenerative outcomes which we must all strive to deliver" (ibid., 25) and identified as follows[5]:

1. Energy is renewable and available to all.
2. Water is responsibly sourced and available to all.
3. Waste does not exist.
4. Natural resources are managed to safeguard communities, animals, and ecosystems.
5. The environment is free from pollution.
6. Our physical presence protects the health of ecosystems and communities.
7. People have the capacity and opportunity to lead fulfilling lives.

Because these properties provide a solid understanding of what a better world might look like, I refer to them henceforth as the seven principles to which companies operating in the state of awakened sustainability will need to adhere. They provide a straightforward narrative based on a clear understanding of ecological and social "boundary conditions within which society can continue to function and evolve, outside of which it cannot" (Broman and Robèrt 2017, 23). They allow businesses to create what the Future-Fit Foundation defines as system value, where "business addresses societal challenges in a holistic way, while not hindering progress toward a flourishing future" (Future-Fit Foundation 2020, 12). Furthermore, these principles are not overly general, like the FSSD sustainability principles, although they remain broad enough to allow for different courses of action without designating a specific pathway to be followed. One example of a course of action based on these principles is the Future-Fit Business Benchmark. The benchmark translates the seven principles to 23 break-even goals, which are presented as thresholds defining the minimum state that companies need to reach (Future-Fit Foundation 2020). Other courses of action would evolve in time, as the criticality of applying these principles is acknowledged further.

---

[5] In addition to these seven properties, the Future-Fit Foundation provides an eighth one that concerns the socioeconomic drivers required to achieve the other seven properties: "Social norms, global governance and economic growth drive the pursuit of future-fitness" (Future-Fit Foundation 2020, 25).

(2) Context: Social justice and regenerative design

While the seven principles on their own make a compelling case for a bold vision, they lack a context for bringing them to fruition most advantageously. A context is necessary to connect the dots between the principles, so they will come across as a coherent manual for companies' operating systems, rather than as a wish list of isolated principles or a set of boxes to check. The context I offer for "gluing" the principles together effectively consists of social justice and regeneration; i.e., making sure our vision is of a fair and regenerative future.

Although both social justice and regeneration are critical for providing proper contextualization, I find the former most essential in the grand scheme of things. After all, most of the seven principles focus on ensuring the stability of the Earth system, which may lead observers to assign less importance to social considerations. This is certainly true regarding climate change, where the discourse can be highly technical and focused mainly on energy and resource use, with inadequate attention paid to the fact that neither the responsibility for climate change nor its consequences are distributed equally.[6]

While recognition is growing that we should aim to create a safe and just operating space for humanity (Raworth 2017), not just a safe one, the prioritization of "safe" ecological boundaries over "just" ones (affecting social well-being) seems to endure, given that all social systems cannot exist without a healthy environment. Yet we should not overlook that this dependency exists on both ends, as "there can be no ecological sustainability without social justice" (Kuhnhenn et al. 2020, 72). In this regard, Leichenko and O'Brien (2019) note: "without attention to social, economic, and political contexts, including politics and power relationships, an emphasis on technical and behavioral responses is unlikely to contribute to the types of structural and systemic changes needed to address the underlying causes of climate change" (48).

Wright defines *social justice* as follows: "in a socially just society, all people would have broadly equal access to the necessary material and social means to live flourishing lives" (Wright 2010, 8). He suggests that this definition does not concern class inequalities only, "it also condemns

---

[6]For example, the Emissions Gap Report 2020 points out that "the combined emissions of the richest 1 percent of the global population account for more than twice the combined emissions of the poorest 50 percent" (UNEP 2020, XIII).

inequalities based on gender, race, physical disabilities, and any other morally irrelevant attribute which interferes with a person's access to the necessary material and social means to live a flourishing life" (Wright 2010, 11). Using this expansive description will bring equity, inclusion, and power considerations front and center, thus amplifying the voices of the marginalized groups and communities that are usually left behind.

This focus on social justice will ensure that when a principle is considered, e.g., energy is renewable and available for all, the access element is taken as seriously as the renewable one. In addition, having social justice at the center will add important pillars such as power to the problem exploration, thus requiring companies to consider that problems are rooted in power and to question whether the solutions they have in mind disrupt or sustain its balance (Gunn-Wright 2020). The ultimate goal is to ensure that sustainability solutions in business are not separated from social justice, and that the latter will become a primary consideration in defining the former.

Regeneration is also an important lens, as it helps anchor the seven principles in a much stronger understanding of what sustainability should entail. It is built on a growing critique about the ability of the current paradigm of sustainability to achieve meaningful outcomes (e.g., Benson and Craig 2014; González-Márquez and Toledo 2020; Gibbons 2020). This critique has been accompanied by calls for the consideration of a regenerative approach (e.g., Gibbons 2020), which could be more effective in addressing the ecological and social challenges we face.

With roots in ecological design that go back to the early twentieth century, regenerative design could be defined as follows: "a system of technologies and strategies, based on an understanding of the inner working of ecosystems that generates designs that regenerate socio-ecological wholes (i.e., generate anew their inherent capacity for vitality, viability and evolution) rather than deplete their underlying life support systems and resources" (Mang and Reed 2012, 8856). In 2007, Reed created a trajectory of environmentally responsible design, wherein he showed the progress from "business-as-usual" design (compliance) through green and sustainable design, and all the way to regenerative design. He made the case that while sustainable design aimed to do no harm (100% less bad), regenerative design was about engaging in and focusing on "the evolution of the whole of the system of which we are part" (Reed 2007, 676). Daniel Christian Wahl, whose book *Designing*

Regenerative Cultures (2016) presents the foundations for a regenerative approach in detail, adds that "regenerative design creates regenerative cultures capable of continuous learning and transformation in response to, and anticipation of, inevitable change" (70).

Regenerative design has been explored in different contexts (agriculture, architecture, etc.), including the design of companies (e.g., Sanford 2017) and even of the economy as a whole (e.g., Fullerton 2015; Raworth 2017). Some of these efforts have been based on developing principle-based frameworks—for example, Sanford (2017) offers seven first principles that companies need to apply to implement a regenerative approach successfully, and Fullerton (2015) presents eight principles underlying regenerative capitalism. Raworth's approach is somewhat different. She sees regenerative design as a driver pushing companies to transform themselves by reconnecting and co-evolving with natural systems, thus becoming more generous in general, "because only generous design can bring us back below the Doughnut's ecological ceiling" (Raworth 2017, 185). My approach is more aligned with Raworth's, as I see regenerative design more as a framework helping to rachet up companies' ambitions for sustainability. The idea is that regenerative design will help companies be more generous as they approach the above-mentioned principles defining the content and consider them as a means to create "a whole system of mutually beneficial relationships" (Wahl 2016, 69).

Finally, I would like to emphasize two points about the context. First, while regeneration has a strong social component, the criticality of social justice issues and the need to prioritize them to provide the vision with a strong human-centered narrative have informed the decision to define social justice as a primary context alongside regenerative design. Second, while I suggest applying regeneration as a design approach to provide a strong context to the seven principles, I mean to enhance sustainability as a general framework for the vision rather than replacing it. Along these lines, I consider the term *regenerative sustainability*[7] as the best way overall to echo regeneration as the highest level on a sustainability trajectory (Reed 2007) that is necessary to shift the economic landscape and make it regenerative by design (Raworth 2017).

---

[7] Gibbons (2020) describes regenerative sustainability as the next wave of sustainability and suggests that "it calls for humans to live in conscious alignment with living systems' principles of wholeness, change, and relationship, as nature does" (3).

Completing the Puzzle: Awakened Sustainability

After combining all three pieces (mental model, content, and context), we receive the full picture of a new vision for a desired future of "awakened sustainability" (see Fig. 6.2). This vision focuses on the notion of awakening, which could range from a very basic level of waking up from a sleep mode, acknowledging the crises we face and what is needed to respond to them adequately, to a deeper spiritual level of change "that challenges and subverts your framework for understanding the nature of reality" (Garlinger 2019, para. 2). It is also about giving new life to the concept of sustainability to reflect the urgency of the climate crisis, as well as the need to prioritize social justice and apply regenerative design in the formation of sustainable solutions.

The awakened sustainability vision is structured to ensure companies go far enough to create a just and safe operating space for humanity (Raworth 2017), but also that it is not too utopian. If anything, I find it to be more aligned with what Wright (2010) calls a "real utopia," which embraces the tension between the imaginary and the practical and is "grounded in the belief that what is pragmatically possible is not fixed independently of our imaginations, but is itself shaped by our visions" (13). Still, the multidimensional sustainability awakening outlined in this

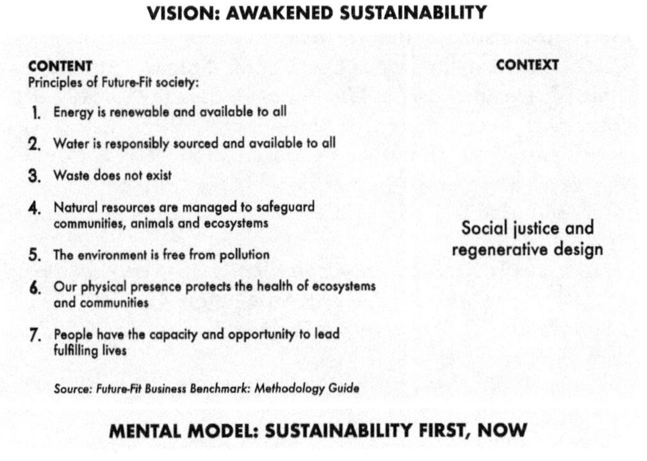

VISION: AWAKENED SUSTAINABILITY

CONTENT
Principles of Future-Fit society:

1. Energy is renewable and available to all

2. Water is responsibly sourced and available to all

3. Waste does not exist

4. Natural resources are managed to safeguard communities, animals and ecosystems

5. The environment is free from pollution

6. Our physical presence protects the health of ecosystems and communities

7. People have the capacity and opportunity to lead fulfilling lives

Source: Future-Fit Business Benchmark: Methodology Guide

CONTEXT

Social justice and regenerative design

MENTAL MODEL: SUSTAINABILITY FIRST, NOW

Fig. 6.2 The mental model, content, and context of the new vision

chapter is easier said than done. It is one thing to articulate where we should be heading, and it is another to actually get there. Therefore, the next step (as part of the backcasting process) will be to consider what needs to be done to realize this vision. This challenge will be the focus of the next two chapters.

REFERENCES

Bakan, Joel. 2020. *The New Corporation: How "Good" Corporations are Bad for Democracy.* New York: Vintage Books.

Benson, Melinda Harm, and Robin Kundis Craig. 2014. "The End of Sustainability." *Society & Natural Resources* 27 (7): 777–82. https://doi.org/10.1080/08941920.2014.901467.

Boulding, J. Russell. 2017. "Peace Culture: An Overview (2000)." In *Elise Boulding: A Pioneer in Peace Research, Peacemaking, Feminism, Future Studies and the Family*, edited by Russell Boulding, 115–20. Switzerland: Springer. https://doi.org/10.1007/978-3-319-31364-1_7.

Broman, Göran Ingvar, and Karl Henrik Robèrt. 2017. "A Framework for Strategic Sustainable Development." *Journal of Cleaner Production* 140 (January): 17–31.

Buis, Alan. 2019. "A Degree of Concern: Why Global Temperatures Matter." *NASA*, June 19. https://rb.gy/e3eyfl.

Dreborg, Karl H. 1996. "Essence of Backcasting." *Futures* 28 (9): 813–28. https://doi.org/10.1016/S0016-3287(96)00044-4.

Ergene, Seray, Subhabrata Bobby Banerjee, and Andrew J. Hoffman. 2020. "(Un)Sustainability and Organization Studies: Towards a Radical Engagement." *Organization Studies.* https://doi.org/10.1177/0170840620937892.

Figueres, Christiana, and Tom Rivett-Carnac. 2020. *The Future We Choose: Surviving the Climate Crisis.* New York: Alfred A. Knopf.

Fletcher, Kate, and Mathilda Tham. 2019. *Earth Logic Fashion Action Research Plan.* London.

Fullerton, John. 2015. *Regenerative Capitalism: How Universal Principles and Patterns Will Shape Our New Economy.* Stonington, CT.

Future-Fit Foundation. 2020. *Future-Fit Business Benchmark: Methodology Guide.* https://rb.gy/l2kp3w.

Garlinger, Patrick Paul. 2019. "What Does It Mean to 'Awaken'?" *Medium*, January 1. https://rb.gy/7cqhnq.

Gibbons, Leah V. 2020. "Regenerative—The New Sustainable?" *Sustainability* 12 (13): 5483. https://doi.org/10.3390/su12135483.

González-Márquez, Iván, and Víctor M. Toledo. 2020. "Sustainability Science: A Paradigm in Crisis?" *Sustainability* 12 (7): 2802. https://doi.org/10.3390/su12072802.

Gunn-Wright, Rhiana. 2020. "Policies and Principles of a Green New Deal." In *Winning the Green New Deal: Why We Must, How We Can*, edited by Varshini Prakash and Guido Girgenti. New York, NY: Simon & Schuster.

Guterres, António. 2018. "Secretary-General's Remarks at the Opening of the COP 24." United Nations Secretary-General. 2018. http://bit.ly/2Cy2zE1.

Holmberg, John. 1998. "Backcasting: A Natural Step in Operationalising Sustainable Development." *Greener Management International* 23: 30–51.

IPCC. 2018. "Global Warming of 1.5°C." https://www.ipcc.ch/sr15/.

———. 2019. "Climate Change and Land." https://www.ipcc.ch/report/srccl/.

Kendall, Geoff, and Martin Rich. 2019. "The Future-Fit Business Benchmark." In *Sustainable Development Goals*, edited by Julia Walker, Alma Pekmezovic, and Gordon Walker, 235–52. Wiley Online Books. https://doi.org/10.1002/9781119541851.ch13.

Kuhnhenn, Kai, Luis Costa, Eva Mahnke, Linda Schneider, and Steffen Lange. 2020. "A Societal Transformation Scenario for Staying Below 1.5°C." https://rb.gy/ld9ody.

Leichenko, Robin, and Karen O'Brien. 2019. *Climate and Society: Transforming the Future*. Cambridge, UK: Polity Press.

Lockton, Dan, and Stuart Candy. 2018. "A Vocabulary for Visions in Designing for Transitions." In *Design as a Catalyst for Change—DRS International Conference 2018*, edited by C. Storni, K. Leahy, M. McMahon, P. Lloyd, and E. Bohemia. Limerick, Ireland. https://doi.org/10.21606/drs.2018.558.

Mang, Pamela, and Bill Reed. 2012. "Regenerative Development and Design." In *Encyclopedia Sustainability Science & Technology*, edited by Robert A. Meyers, 2nd Editio, 8855–79. New York, NY: Springer. https://doi.org/10.1007/978-1-4419-0851-3_303.

McKibben, Bill. 2017. "Winning Slowly is the Same as Losing." *Rolling Stone*, December. https://rol.st/2NLguKh.

Meadows, Donella H., Dennis L. Meadows, and Jorgen Randers. 1992. *Beyond the Limits: Confronting Global Collapse, Envisioning a Sustainable Future*. Post Mills, Vermont: Chelsea Green Publishing.

Missimer, Merlina, Karl-Henrik Robèrt, Göran Broman, and Harald Sverdrup. 2010. "Exploring the Possibility of a Systematic and Generic Approach to Social Sustainability." *Journal of Cleaner Production* 18 (10–11): 1107–12. /https://doi.org/10.1016/j.jclepro.2010.02.024.

Mumford, Michael D., and Jill M. Strange. 2002. "Vision and Mental Models: The Case of Charismatic and Ideological Leadership." In *Transformational and Charismatic Leadership: The Road Ahead*, edited by B. J. Avolio and F. J. Yammarino, 109–142. Oxford, UK: Elsevier.

Ny, Henrik, Jamie P. MacDonald, Göran Broman, Ryoichi Yamamoto, and Karl-Henrik Robért. 2008. "Sustainability Constraints as System Boundaries: An Approach to Making Life-Cycle Management Strategic." *Journal of Industrial Ecology* 10 (1–2): 61–77. https://doi.org/10.1162/108819806775545349.

Ocasio-Cortez, Alexandria. 2019. *H.Res.109—Recognizing the Duty of the Federal Government to Create a Green New Deal*. Washington, DC: 116th Congress (2019–2020). https://www.congress.gov/bill/116th-congress/house-resolution/109/text.

Quist, J. 2007. *Backcasting for a Sustainable Future: The Impact After 10 Years*. Delft, the Netherlands: Eburon Academic Publishers.

Raworth, Kate. 2017. *Doughnut Economics: Seven Ways to Think Like a 21st-Century Economist*. London: Random House.

Reed, Bill. 2007. "Shifting From 'Sustainability' to Regeneration." *Building Research & Information* 35 (6): 674–80. https://doi.org/10.1080/096132 10701475753.

Robèrt, Karl-Henrik. 2002. *The Natural Step Story : Seeding a Quiet Revolution*. Gabriola Island, Canada: New Society Publishers.

Robinson, John B. 1990. "Futures Under Glass: A Recipe for People Who Hate to Predict." *Futures* 22 (8): 820–42. https://doi.org/10.1016/0016-3287(90)90018-D.

Sanford, Carol. 2017. *The Regenerative Business: Redesign Work, Cultivate Human Potential, Achieve Extraordinary Outcomes*. New York, NY: Nicholas Brealey.

Seinfeld, Jerry, Larry David, and Andy Cowan. 1994. "Seinfeld: The Opposite." May 19, 1994. https://rb.gy/zvjhh2.

Senge, Peter M. 1990. *The Fifth Discipline: The Art & Practice of the Learning Organization*. New York: Doubleday.

Strange, Jill M., and Michael D. Mumford. 2005. "The Origins of Vision: Effects of Reflection, Models, and Analysis." *The Leadership Quarterly* 16 (1): 121–48. https://doi.org/10.1016/j.leaqua.2004.07.006.

UNEP. 2020. *Emissions Gap Report 2020*. Nairobi. https://rb.gy/ehhfom.

Vergragt, Philip J., and Jaco Quist. 2011. "Backcasting for Sustainability: Introduction to the Special Issue." *Technological Forecasting and Social Change* 78 (5): 747–55.

Wahl, Daniel Christian. 2016. *Designing Regenerative Cultures*. Axminster, UK: Triarchy Press.

Wright, Erik Olin. 2010. *Envisioning Real Utopias*. London, UK: Verso.

CHAPTER 7

What Needs to Be True?

Abstract This chapter presents the first part of a roadmap aimed at realizing the shift from sustainability-as-usual to awakened sustainability. Applying inversion, I consider what needs to be true in the world to allow for this shift to take place. Lessig's "New Chicago School" framework, which focuses on four forces (conditions) regulating behavior (law, markets, social norms, and architecture) provides a useful framework for this purpose, allowing us to give proper weight to the institutional environment's impact on the behavior of companies. Building on existing research, the chapter lays out key changes in each condition that need to occur to make awakened sustainability possible: changes in corporate law (law), financial incentives (markets), social norms and meanings (social norms), and organizational culture (architecture).

Keywords Social norms · Organizational culture · Green bond · Chicago new school · Corporate law · Green credit

Design mode means the outcome of combining three human gifts: critical sense (the ability to look at the state of things and recognize what cannot, or should not be, acceptable), creativity (the ability to imagine something that does not yet exist), and practical sense (the ability to recognize feasible ways of getting things to happen). (Manzini 2015, 31)

© The Author(s), under exclusive license to Springer Nature   117
Switzerland AG 2021
R. Godelnik, *Rethinking Corporate Sustainability in the Era
of Climate Crisis*, https://doi.org/10.1007/978-3-030-77318-2_7

After laying out a vision for a desired state of awakened sustainability in the last chapter, it is time to shift our focus to the last and critical part of the design mode (see the quote above)—realizing the vision of awakened sustainability. The exploration of what is needed to make this vision a reality will be conducted in two steps: (1) considering what needs to be true in the world for awakened sustainability to take place, and (2) figuring out how to actualize step 1.

Presented in this chapter, step no. 1 applies inversion, which "allows us to flip the problem around and think backward" (Parrish and Beaubien 2019, 143) to start exploring what is needed to realize awakened sustainability. As Parrish and Beaubien (2019) point out, inversion can be applied in two ways: "1) Start by assuming that what you're trying to prove is either true or false, then show what else would have to be true. 2) Instead of aiming directly for your goal, think deeply about what you want to avoid and then see what options are left over" (143). In this case, I apply the first option, asking: "If the vision of awakened sustainability is indeed true, then what else (i.e., what conditions) in the world would need to be true?"

Not only does this approach have a long history of applications, including Edward Bernays' famous campaign to increase the sales of cigarettes to women almost a century ago (Axelrod 2008; Parrish and Beaubien 2019),[1] but it is also strategic in nature. As Martin (2020) suggests, the question "what would have to be true?" is the most important one in strategy, given how it reframes our thinking and helps us replace preconceived biases with exploration.[2] The second step, which will be presented in the next chapter, will focus on the "how" question: i.e., how to make the conditions that need to be true actually true.

Our first challenge is to define what conditions to focus on. What is it exactly that needs to be true in the world to allow for the desired vision of awakened sustainability to be materialized? To do so, I draw on Lessig's (1998) "New Chicago School" model,[3] which asserts that behavior is

---

[1] Parrish and Beaubien report that "Bernays did not ask, 'How do I sell more cigarettes to women?' Instead, he wondered, if women bought and smoked cigarettes, what else would have to be true? What would have to change in the world to make smoking desirable to women and socially acceptable?" (Parrish and Beaubien 2019, 144–45).

[2] This question is a key part in Martin's (2020) strategic choice structuring process (an initial version of this process can be found in Martin [1997]).

[3] This model is also referred to as "the pathetic dot theory" (Lessig 2006).

regulated by four constraints: law, markets, social norms, and architecture, which are all connected to one another. While Lessig presented his model in a regulatory context (i.e., the consideration of forces outside of law, such as markets, social norms, and architecture, which play an important role in regulating people's behavior), I find it useful to build on this model in the context of corporate sustainability as well.[4]

Drawing on Lessig's model recognizes the complexity of change-making in organizations, especially when it is not an incremental deviation from the status quo, but more of a transformative change. Avoiding failure, which is the outcome of most change initiatives in organizations (Keller and Schaninger 2019), requires an approach that is more holistic in nature. In that sense, Lessig's approach seems like a good fit. First, by considering laws and social norms as tools to shape behavior, it gives sufficient weight to the institutional environment's impact on the behavior of companies, which includes "rules, norms, and beliefs surrounding economic activity that define or enforce socially acceptable economic behavior" (Oliver 1997, 698). This approach is in line with a large body of literature showing how corporate action, including on environmental and social issues, is shaped by institutions (e.g., Campbell 2007; Bansal 2005; Hoffman 2001).[5]

The multiplicity of Lessig's model also echoes the comprehensiveness of multidimensional approaches to explaining institutions, such as Scott's (1995) three pillars (regulative, normative, and cognitive) and DiMaggio and Powell's (1983) three mechanisms of institutional isomorphic change (coercive, normative, and mimetic). In addition, the final two constraints (markets and architecture) help form a more polycentric framework that go beyond putting all of the eggs in one "institutional basket"—i.e., taking into account that the ways in which organizations behave and change depends on more than just regulation and normative pressures for conformity. The market's constraint sheds light on the role of economic forces and conditions in shaping companies' behavior.

---

[4] Lessig's model has been used in research exploring regulatory contexts, including ones concerning cyberspace, which was a focus of Lessig (e.g., Lessig 2006). It was also employed in the context of corporate sustainability by Sjåfjell and Taylor (2019), who developed a "'regulatory ecology' of corporate purpose, exploring the interaction of company law, with social norms, such as shareholder primacy and sustainability" (41–42).

[5] Additional resources on the impact of institutional environment on organizational behavior can be found in Swaminathan and Wade (2016).

The final constraint, architecture, regards how things (from buildings to cyberspace) are regulated by their design, suggesting that "features of the world— whether made, or found —restrict and enable in a way that directs or affects behavior" (663). This constraint helps us zoom in on the organizational context, i.e., how the design of the organization regulates its behavior.

Furthermore, drawing on Lessig's model is also aligned with taking a strategic design approach, which pays close attention to what Hill (2012) calls "dark matter." Hill suggests this metaphor to represent the important invisible elements responsible for the value created by organizations. According to Hill, dark matter in this sense refers to "organisational culture, policy environments, market mechanisms, legislation, finance models and other incentives, governance structures, tradition and habits, local culture and national identity, the habitats, situations and events that decisions are produced within." Similar to dark matter, these elements are out of sight, are difficult to trace, and may get less attention; nevertheless, they are important for understanding the matter that is visible (i.e., the products and services produced by companies). Lessig's model focuses on and engages with the dark matter shaping corporate sustainability, allowing us to see the whole picture more clearly and to identify how to produce eventually effective interventions that are "systemic, permanent, influential" (Hill 2012).

I will look now in further detail into what needs to be true for each one of the four conditions (law, social norms, markets, architecture) to allow awakened sustainability to take place. While this challenge could be excruciating, given the large number of propositions to choose from, it is also an invitation to consider what elements within each condition are significant enough to support and foster a transformational change, such as the one that the vision of awakened sustainability entails. The changes presented for the four conditions (see Fig. 7.1) are by no means the only options to consider; instead, they offer examples of the changes that need to take place inside and outside of companies to support a new state of awakened sustainability.

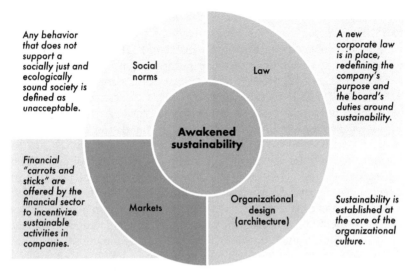

Based on Lessig's (1998) "New Chicago School" framework

Fig. 7.1 What needs to be true for each of the four conditions supporting awakened sustainability

CONDITION #1: LAW—A NEW CORPORATE LAW, REDEFINING THE COMPANY'S PURPOSE AND THE BOARD'S DUTIES AROUND SUSTAINABILITY

Mayer (2020) suggests that "since the corporation is a product of the law, the law can define the nature of the corporation" (7). The law indeed plays an important role in shaping organizational behavior, providing formal systems for different actors to act on and impact one another (Hoffman 2001). To support awakened sustainability properly, we need laws in place that clearly direct companies to put sustainability first. Hence, I suggest that we need a corporate law that includes purpose as a legal concept and considers sustainability as the main purpose of the company.

This suggestion is largely in line with similar recommendations from the SMART Project[6] and the British Academy's Future of the Corporation program (henceforth, the British Academy). The SMART project's researchers suggest defining corporate purpose legally in general terms, as "creating sustainable value within planetary boundaries" (Sjåfjell et al. 2019). Sjåfjell (2020) explains that sustainable value "must reflect the multifaceted and interconnected environmental, social, cultural, economic and governance aspects of securing the social foundation for humanity" (5). The British Academy's researchers came up with a similar idea to reformulate corporate law around a purpose defined as follows: "The purpose of business is to profitably solve the problems of people and planet, and not profit from causing problems" (British Academy 2019, 8). Both groups of researchers also want to redefine the board's duties (and board members' fiduciary duty) to ensure they are aligned with and accountable for the new purpose. Reyes (2020) recommends a similar measure, adding that to avoid circumvention of it, board reform should include the right of workers to elect up to half of the board members (similarly to the requirement for German companies with more than 2000 employees).

I concur with these recommendations, with an emphasis on ensuring that the definition of corporate purpose provides a clear understanding of what sustainability stands for. While general in nature, the purpose should stand on solid environmental and social foundations and be explicit about the need of companies to act with urgency. The SMART project's references to planetary boundaries and to Raworth's (2017) social foundation of the "doughnut" framework are good examples for the first two parameters (environmental and social foundations, respectively). Including a commitment to align the company's actions with the latest recommendations of the Intergovernmental Panel on Climate Change (IPCC) on a timeline for action (see, e.g., IPCC 2018) would add the third parameter (urgency) to the mix.

Additionally, it should be clear that the reformulation of the law substantially changes the power dynamics between sustainability and profitmaking. Not only does it make it clear that shareholder primacy is obsolete, but it also goes far beyond attempts to level up stakeholders'

---

[6] SMART stands for Sustainable Market Actors and Responsible Trade. See more details about the project at https://cordis.europa.eu/project/id/693642.

considerations, including hybrid legal structures (e.g., Benefit Corporations[7]) and suggestions that new corporate charters will require large companies to consider the interests of all stakeholders in their decisions (see, e.g., Warren 2018). As Stout (2012) points out, "maximizing shareholder value is not a managerial obligation, it is a managerial choice" (32). Thus, having a new corporate law in place that prioritizes sustainability as the company's purpose should help make it clear for managers that the purpose of the business and the duties of its board are to improve the health of society and the planet, not to maximize shareholder returns.

CONDITION #2: SOCIAL NORMS—DEFINING ANY BEHAVIOR THAT DOES NOT SUPPORT A SOCIALLY JUST AND ECOLOGICALLY SOUND SOCIETY AS UNACCEPTABLE

In two books that share the same title (*How Change Happens*), both authors emphasize the effectiveness of social norms as a key change enabler. Leslie Crutchfield (2018) suggests that "savvy social change makers understand that, if they want to achieve impact on their issues, they must shift social norms, not just reform policies and laws" (82). Cass Sunstein (2019) adds that "when social norms change, social meaning changes too. Thus the social meanings of lighting up a cigarette, engaging in an act of sexual harassment, using a condom, or refusing to eat meat are very different now from what they were in 1961 because of dramatic shifts in underlying norms" (39).

Social norms[8] have played an important role in shaping environmental behavior changes in organizations in the past (e.g., Bansal and Roth 2000; Hoffman 2001). They are also very likely to have a key role in any transition toward a sustainable future. This is true in particular with regard to climate change: "changes in social norms among the general public and senior decision makers in government, industry and business will be critical for achieving net zero goals intended to limit warming below 1.5–2°C" (Brooks 2020, 14). My focus is more specifically the impact of social norms among the general public on how companies conduct business.

---

[7] See at https://benefitcorp.net/what-is-a-benefit-corporation.

[8] The concept of social norms is presented in Chapter 1, including Posner's (1997) definition: "rule that is neither promulgated by an official source, such as a court or a legislature, nor enforced by the threat of legal sanctions, yet is regularly complied with" (365).

Having a state of awakened sustainability in place will require social norms that support the transformative changes the new state entails. The shift to a new state will necessitate a reexamination of what is socially acceptable and what is not, defining any behavior that does not support a socially just and ecologically sound society as unacceptable and vice versa.

The new social norms should contest and eventually replace the current ones, which permit many unsustainable corporate practices, from using fossil fuel and creating waste to not paying workers fair wages and contributing to social injustices around the world. The new norms could be supported by norm entrepreneurs who work actively to challenge existing norms and replace them with new ones (Sunstein 1996, 2019). They, for example, could help redefine what is acceptable and expected in regard to potentially ambiguous concepts such as fairness and inclusion,[9] which may be interpreted opportunistically by companies. Another course of action is to change the meaning of social license to operate, contesting the license of certain companies and industries to operate: for example, those not acting with urgency to become fossil free.

"Flipping" dominant social norms from acceptable to unacceptable would result in creating alternative new social meanings—for example, by "reframing the idea of progress around human wellbeing" (Brooks 2020, 17). New social meanings will allow for a greater regulatory effect of the new norms, because as Lessig (1998) explains "the cost (whether internal or external) of deviating from a social norm is not constituted by the mere deviation from a certain behavior; it is a cost in part constituted by the meaning of deviating from a certain behavior" (680). Thus, new meanings will make a clearer understanding of corporate practices possible, along with how they should be perceived in the context of awakened sustainability, which could lead to greater conformity with the new state.

---

[9] Take for example the movement to establish $15 as the minimum fair wage in the U.S. (https://fightfor15.org/), which can be considered as an effort to articulate to companies what can be considered as a fair pay.

CONDITION #3: MARKETS—FINANCIAL CARROTS AND STICKS OFFERED BY THE FINANCIAL SECTOR TO INCENTIVIZE SUSTAINABLE ACTIVITIES IN COMPANIES

Transforming the financial system is key to fighting climate change and securing a future shaped by the vision of awakened sustainability. So far this system has been much more part of the problem than the solution, by continuing to support fossil fuel projects and carbon-intensive industries, as well as by paying little to no attention to climate risks (Green et al. 2020). Mark Carney (2015), a former Bank of England governor, suggested that these failures represent "the tragedy of the horizon"; i.e., the inherent gap between the short-term mindset of the financial sector and the long-term nature of climate change, which leads the former to ignore the risks of the latter. This gap results in "imposing a cost on future generations that the current generation has no direct incentive to fix" (4). The tragedy is that "once climate change becomes a defining issue for financial stability, it may already be too late" (4).

Avoiding this tragedy, Carney suggests, requires designing a new horizon, to realign the financial sector with the climate crisis. Doing so, according to Carney, would necessitate changes in three critical areas: (1) reporting; (2) risk; and (3) return (Carney 2019a). I find Carney's framework useful, not only "to bring climate risks and resilience into the heart of financial decision making" (Carney 2019b, 13), but to go even further and help prioritize sustainability over other considerations as suggested by the awakened sustainability approach. Following Lessig's (1998) point that markets regulate through prices, I suggest that proper consideration of these areas, especially risk and return,[10] by the financial sector can provide clear price signals to companies on the direction they should take.

In both the areas of return and risk, the financial sector needs to create more effective carrots and sticks to back the state of awakened sustainability. In the case of risk, capital providers, such as banks, asset managers, and insurers need not only to understand and manage climate risks better, by getting companies to improve their governance, management, and disclosure of them (Carney 2019b), but they also need to take these risks into account in their offerings. Take insurers for example. According to Waddell et al. (2020), insurers could take action on climate in a

---

[10] The first area of focus (reporting) is also important, but it has more of a supportive role to the other two.

number of areas, including their underwriting portfolios, driving changes in their customer base to incentivize the pursuit of lower carbon footprints. Waddell et al. (2020) suggest that one way to do so is to exclude from the insurance company's underwriting portfolio any clients involved with fossil fuel projects or having high carbon footprints. Another way is to adjust the premiums based on the client's level of action on climate, whereby more action results in lower premiums and vice versa.

As for return optimization considerations, the financial sector should focus on recognizing that addressing challenges like climate change can create financially rewarding opportunities for investors. This realization should result in mainstreaming sustainable investing, which rewards companies pursuing sustainability and penalizes those that do not in terms of access to and cost of capital. To do so, investors need to take on strategies that clearly reflect this "sticks and carrots" approach. One option, according to Reyes (2020), is green credit policies, which can prioritize sustainable projects and companies in receiving loans. Such policies can "establish minimum requirements for the proportion of bank loans targeting 'green' projects and upper limits on lending to carbon-intensive sectors" (Reyes 2020, 26). By providing sustainability-linked loans, for example, banks can influence not just the availability of capital, but also its cost, as the cost of the loan goes down when the borrower meets certain sustainability goals (Mitchell et al. 2020). This measure and others, such as issuing green or climate bonds, (which could be part of green quantitative easing programs initiated by central banks, see e.g., Reyes 2020) should be supported by clear criteria as to what sustainability stands for and, accordingly, which projects and activities should be supported by the financial system.[11]

Condition #4: Architecture—Establishing Sustainability at the Core of the Organizational Culture

Similarly to the design of roads, buildings, or software, the design of an organization could act as an architectural constraint, regulating the behavior and actions of the people operating in it. In this case, organizational culture would be a good fit to represent the architectural

---

[11] There are some initial efforts to create such systems for the financial sector, including a taxonomy developed by the EU (Reyes 2020).

constraint in Lessig's model, given its impact on how organizations operate (Alvesson and Sveningsson 2008), including as pertains to corporate sustainability. As Baumgartner (2009) notes: "one important point for companies willing to be more sustainable is the awareness of their organizational culture and to reach a fit between the culture and the sustainability activities" (112). Using organizational culture will also allow to consider the role of corporate leaders, who are key to creating and managing culture in organizations (Schein 2004).

To assess what the organizational culture should look like in order to give adequate support to the state of awakened sustainability I will build on Bertels et al.'s (2010) extensive research on how to build a culture of sustainability. First, we need to define what such a culture stands for in the context of awakened sustainability. Adjusting Bertels et al.'s definition to this context, I suggest that a culture of sustainability is one where the members of the organization hold shared assumptions and beliefs about the need to put sustainability first.[12] According to Bertels et al. (2010), the organizational practices needed to build a culture of sustainability can be separated into two groups—informal and formal. The former aims "to establish and reinforce shared values and shared ways of doing things that align the organization with its journey to sustainability" (13), while the latter seeks "to guide behavior through the rules, systems, and procedures" (13). This framework seems to align with Schein's (2004) popular framework of three levels of culture (artifacts, espoused values, and underlying assumptions), whereby the formal approach is aligned with the "artifacts" level and the informal approach is aligned with the "espoused values" level. In addition, it coincides with Graham et al.'s (2016) research suggesting that the effectiveness of culture in organizations depends on an alignment in the organization between values and norms and formal institutions.

The informal approach presented by Bertels et al. (2010) includes a number of practices that can help build and support awakened sustainability. One such practice that needs to be in place is to "model," whereby the company's leadership shows it walks the talk by prioritizing sustainability in discussions and decision-making and by providing clear signals for the entire organization about their support for sustainability. The key,

---

[12] Bertels et al. (2010) define a culture of sustainability as "one in which organizational members hold shared assumptions and beliefs about the importance of balancing economic efficiency, social equity and environmental accountability" (10).

the researchers explain "is to ensure alignment and consistency between the organization's sustainability goals and the actions of its senior leadership" (22). Another practice is "support," which is about making it easier for employees to make decisions that prioritize sustainability at work as well as in their personal life. The latter is important as it can be an embodiment of the practical sphere in O'Brien and Sygna's (2013) three spheres of transformation, potentially supporting changes in employees' mindsets, and the development of a "sustainability mindset" (Rimanoczy 2021). It could also support another important practice, "frame," which is about (re)framing sustainability in ways that reflect organizational values and priorities. One example would be for the organization to frame it as urgent, while another would be to portray sustainability as being about social justice and well-being.

Organizational processes (i.e., formal practices), which include more traditional avenues, such as development and implementation of sustainability-based policies (procurement, code of conduct, etc.), as well as assigning responsibility to the senior leadership could provide strong signals about the direction the organization is taking. Linking compensation to sustainability performance could also help focus employees' attention on awakened sustainability. To do so, for example, the main part of executive bonuses and short-term/long-term incentive plans will be based on metrics tied to awakened sustainability principles. According to Hong et al. (2015), using such direct incentives could help enhance the company's sustainability performance. Making sustainability criteria an essential part of promotion, hiring, and firing decisions could further reinforce its importance to the organization.[13] I find it particularly important in terms of hiring. As Bertels et al. (2010) explain, "to build and support a culture of sustainability, an organization's recruiting and selection processes should strategically build a pool of human capital with values and skills that support the journey toward sustainability" (25).

As Lessig (1998) notes, the four constraints operate together. The abovementioned four conditions do the same, as they operate jointly, influencing one another. For example, the power of new social norms can enable the legislation of new corporate laws, which in turn can support changes in the markets, which can then help organizational culture evolve

---

[13] Bertels et al. (2010) refer to hiring and promotion as informal practices, but I find that they fit the definition of formal practices better, especially when codified as formal processes. This approach also accords with that of Graham et al. (2016).

in a more desired direction, and so on. In this case, given the complexity of the shift from sustainability-as-usual to awakened sustainability, along with the multifaceted nature of corporate sustainability, it is even more important to see that all four conditions must occur in tandem for awakened sustainability to be possible. Recognizing the necessity of these "dark matter" conditions for realizing a new model of corporate sustainability echoes Crutchfield's (2018) notion of the need for changing both minds and policy to achieve lasting systemic change, as well as a more holistic view of the organization in general. However, the question remains of how to ensure that the required changes in these conditions take place. This will be the focus of the next and final chapter.

REFERENCES

Alvesson, Mats, and Stefan Sveningsson. 2008. *Changing Organizational Culture*. London: Routledge.

Axelrod, Alan. 2008. *Profiles in Folly: History's Worst Decisions and Why They Went Wrong*. New York, NY: Sterling Publishing.

Bansal, Pratima. 2005. "Evolving Sustainably: A Longitudinal Study of Corporate Sustainable Development." *Strategic Management Journal* 26 (3): 197–218.

Bansal, Pratima, and Kendall Roth. 2000. "Why Companies Go Green: A Model of Ecological Responsiveness." *Academy of Management Journal* 43 (4): 717–36. https://doi.org/10.2307/1556363.

Baumgartner, Rupert J. 2009. "Organizational Culture and Leadership: Preconditions for the Development of a Sustainable Corporation." *Sustainable Development* 17 (2): 102–13. https://doi.org/10.1002/sd.405.

Bertels, Stephanie, Lisa Papania, and Daniel Papania. 2010. "Embedding Sustainability in Organizational Culture: A Systematic Review of the Body of Knowledge for Business Sustainability in Organizational Culture." *Network for Business Sustainability*.

British Academy. 2019. "Principles for Purposeful Business." https://rb.gy/dhd2ox.

Brooks, Nick. 2020. "Shifting Norms and Values for Transitions to Net-Zero." Working Paper, Climate-KIC (WP).

Campbell, John L. 2007. "Why Would Corporations Behave in Socially Responsible Ways? An Institutional Theory of Corporate Social Responsibility." *Academy of Management Review* 32 (3): 946–67. https://doi.org/10.5465/amr.2007.25275684.

Carney, Mark. 2015. "Breaking the Tragedy of the Horizon." *Bank of England*. https://rb.gy/wbsplk.

———. 2019a. "A New Horizon." *SUERF Policy Note*, no. 65. https://rb.gy/gpotop.

———. 2019b. "Fifty Shades of Green." *IMF Finance and Development*, December. https://rb.gy/orrkre.

Crutchfield, Leslie R. 2018. *How Change Happens: Why Some Social Movements Succeed While Others Don't*. Hoboken, NJ: Wiley.

DiMaggio, Paul J., and Walter W. Powell. 1983. "The Iron Cage Revisited: Institutional Isomorphism and Collective Rationality in Organizational Fields." *American Sociological Review* 48 (2): 147–60. https://doi.org/10.2307/2095101.

Graham, John R., Campbell R. Harvey, Shivaram Rajgopal, and Shivaram Rajgopal. 2016. "Corporate Culture: Evidence from the Field." *SSRN Electronic Journal*. http://www.ssrn.com/abstract=2805602.

Green, Andy, Gregg Gelzinis, and Alexandra Thornton. 2020. "Financial Markets and Regulators Are Still in the Dark on Climate Change." *Center for American Progress*, June 29. https://rb.gy/papecw.

Hill, Dan. 2012. *Dark Matter and Trojan Horses: A Strategic Design Vocabulary*. Moscow: Strelka Press.

Hoffman, Andrew J. 2001. *From Heresy to Dogma: An Institutional History of Corporate Environmentalism*. Palo Alto, CA: Stanford University Press.

Hong, Citation, Zhichuan Li, and Dylan B. Minor. 2015. "Corporate Governance and Executive Compensation for Corporate Social Responsibility." Harvard Business School Working Paper Series # 16–014. http://nrs.harvard.edu/urn-3:HUL.InstRepos:19052357.

IPCC. 2018. "Global Warming of 1.5 °C." https://www.ipcc.ch/sr15/.

Keller, Scott, and Bill Schaninger. 2019. *Beyond Performance 2.0: A Proven Approach to Leading Large-Scale Change*. 2nd ed. Hoboken, NJ: Wiley.

Lessig, Lawrence. 1998. "The New Chicago School." *The Journal of Legal Studies* 27 (S2): 661–91. https://doi.org/10.1086/468039.

———. 2006. *Code: Version 2.0*. New York: Basic Books.

Manzini, Ezio. 2015. *Design, When Everybody Designs: An Introduction to Design for Social Innovation*. The MIT Press.

Martin, Roger. 1997. "Strategic Choice Structuring." White Paper. https://rb.gy/5nvaps.

———. 2020. "Strategy & Design Thinking." *Medium*, December 14. https://rb.gy/ycj18k.

Mayer, Colin. 2020. "The Future of the Corporation and the Economics of Purpose." Finance Working Paper No. 710/2020. European Corporate Governance Institute. https://doi.org/10.2139/ssrn.3731539.

Mitchell, James, Lindsey Schafferer, Tyeler Matsuo, and Radhika Lalit. 2020. "Breaking the Code: Deciphering Climate Action Efforts in the Financial

Sector." *Rocky Mountain Institute*. https://rmi.org/insight/breaking-the-code/.

O'Brien, Karen, and Linda Sygna. 2013. "Responding to Climate Change: The Three Spheres of Transformation." In *Proceedings of Transformation in Changing Climate International Conference*, 16–23. Oslo, Norway: University of Oslo.

Oliver, Christine. 1997. "Sustainable Competitive Advantage: Combining Institutional and Resource-Based Views." *Strategic Management Journal* 18 (9): 697–713.

Parrish, Shane, and Rhiannon Beaubien. 2019. *The Great Mental Models Volume 1: General Thinking Concepts*. Ottawa, Canada: Latticework Publishing Inc.

Posner, Richard A. 1997. "Social Norms and the Law: An Economic Approach." *American Economic Review* 87 (2): 365–69. https://doi.org/10.2307/295 0947.

Raworth, Kate. 2017. *Doughnut Economics: Seven Ways to Think Like a 21st-Century Economist*. London: Random House.

Reyes, Oscar. 2020. *Change Finance, Not the Climate*. Amsterdam and Washington, DC: Transnational Institute (TNI) and the Institute for Policy Studies (IPS).

Rimanoczy, Isabel. 2021. *The Sustainability Mindset Principles: A Guide to Develop a Mindset for a Better World*. New York, NY: Routledge.

Schein, Edgar H. 2004. *Organizational Culture and Leadership*. 3rd ed. San Francisco: Jossey-Bass.

Scott, William Richard. 1995. *Institutions and Organizations*. Thousand Oaks, CA: Sage.

Sjåfjell, Beate. 2020. "Sustainable Value Creation Within Planetary Boundaries-Reforming Corporate Purpose and Duties of the Corporate Board." *Sustainability (Switzerland)* 12 (15): 6245. https://doi.org/10.3390/SU1215 6245.

Sjåfjell, Beate, Jukka Mähönen, Mark B. Taylor, Eléonore Maitre-Ekern, Maja van der Velden, Tonia Novitz, Clair Gammage, Jay Cullen, Marta Andhov, and Roberto Caranta. 2019. "Supporting the Transition to Sustainability: SMART Reform Proposals."

Sjåfjell, Beate, and Mark B. Taylor. 2019. "Clash of Norms: Shareholder Primacy vs. Sustainable Corporate Purpose." *International and Comparative Corporate Law Journal* 13 (3): 40–66. https://doi.org/10.2139/ssrn.3444050.

Stout, Lynn. 2012. *The Shareholder Value Myth: How Putting Shareholders First Harms Investors, Corporations, and the Public*. San Francisco: Berrett-Koehler Publishers.

Sunstein, Cass R. 1996. "Social Norms and Social Roles." *Columbia Law Review* 96 (4): 903–68. https://doi.org/10.2307/1123430.

———. 2019. *How Change Happens*. Cambridge, MA: The MIT Press.

Swaminathan, Anand, and James B. Wade. 2016. "Institutional Environment." In *The Palgrave Encyclopedia of Strategic Management*, edited by Mie Augier and David J. Teece, 1–7. London: Palgrave Macmillan. https://doi.org/10. 1057/978-1-349-94848-2_608-1.

Waddell, Rebecca, Douglas Beal, and David Cockerill. 2020. "Insurers Take Up the Climate Fight." *BCG*, August 19. https://rb.gy/d7pjaz.

Warren, Elizabeth. 2018. *S.3348—Accountable Capitalism Act*. 115th Congress (2017–2018). https://rb.gy/huk35s.

CHAPTER 8

Breaking on Through to the Other Side: How to Make Change Happen?

Abstract This chapter presents the second part of the roadmap for realizing the shift from sustainability-as-usual to awakened sustainability. It explores how to make the changes in the four conditions (law, markets, social norms, and organizational design) that are key to achieving awakened sustainability. The chapter draws on the importance of mental models in system transformation, suggesting that the necessary changes in the four conditions will need to be supported by a shift in the mental model dominating corporate sustainability, from "shareholder capitalism 2.0" to "sustainability first, NOW." It proceeds with a detailed theory of change built on the power of narratives of transformative change to win the "battle of narratives," which would result eventually in the desired changes in the mental model and consequently in the four conditions. Finally, the shift to awakened sustainability is connected to the different transformation pathways presented in Chapter 5.

Keywords Theory of change · Narratives · Overton window · Political alignment · 3.5% rule · Three horizons

R. Godelnik, *Rethinking Corporate Sustainability in the Era of Climate Crisis*, https://doi.org/10.1007/978-3-030-77318-2_8

To generate equitable and sustainable results, we have to design our responses in such a way that we address the whole culture and system—a conscious full-spectrum response. (Sharma 2017, 5)

This final chapter focuses on connecting the dots between the different ideas and themes that the second part of the book has so far shared, providing a course of action to enable the changes in the four conditions necessary to move from sustainability-as-usual to awakened sustainability.[1] I find this course of action to be essential, as it helps strengthen the notion of awakened sustainability as a "real utopia" (Wright 2010), or in other words an achievable alternative to sustainability-as-usual.

CHANGING THE MENTAL MODEL

The previous chapter established the four conditions that are key to achieving awakened sustainability (law, social norms, markets, and organizational design) and what would have to be true for each one of them—that is, what changes they would need to go through—to bring awakened sustainability into being. Our focus now shifts to figuring out how we might make these changes happen. To do so, we need to approach this challenge from a holistic and systemic point of view, which considers among other things the broader context in which all of this happens (i.e., the recent key trends and events detailed in Chapter 5), and in particular the climate crisis, given its critical role in shaping this decade and beyond.

With this context in mind, I find it helpful to apply a multilevel approach to systems change, which will consider law, social norms, markets, and organizational design as interdependent system conditions that can operate and drive change on different levels. Kania et al.'s (2018) model of six conditions of systems change is an example of this approach, examining "six interdependent conditions that play significant roles in holding a social or environmental problem in place" (3) and organizing them in three levels "with respect to their visibility and their ability to transform a system" (3).

---

[1] See Chapter 7 for further detail.

The three levels in Kania et al.'s (2018) model are as follows: (1) Structural change, which includes changes that are more visible in policies, practices, and resource flows. (2) Relational change (i.e., changes in dynamics, relationships, and connections), which are less visible. (3) Transformative change; this level, which is the least visible, covers mental models. The authors point out that "the less explicit conditions are the most challenging to clarify but can have huge impacts on shifting the system" (Kania et al. 2018, 5). Furthermore, the authors suggest that the least visible level of mental models is the most impactful one, and while "mental models are not necessarily 'more causative' than other conditions" (5), they are the most important driver of activity in any system. Thus, unless attention is paid to the level of mental models, it will be difficult to sustain changes on the other two levels. This approach is aligned with a broader recognition of the importance of mental models in shaping people's understanding of the world, decision-making, and actions (see, e.g., Senge 1990; World Bank 2015).

Categorizing law, social norms, markets, and organizational design according to the three levels (see Fig. 8.1) suggests that they operate

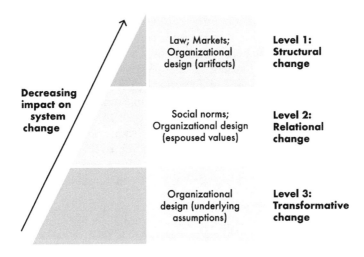

Based of Kania et al.'s (2018) Six Conditions of Systems Change model

Fig. 8.1 The categorization of law, social norms, markets, and organizational design according to Kania et al.'s (2018) three-level model of systems change

mostly on the first and second levels (structural and relational changes, respectively).[2] Utilizing Schein's (2004) framework of three levels of culture in organizations (artifacts, espoused values, and underlying assumptions) to explain the organizational design condition, only "underlying assumptions" operate on the transformative change level, acting as the "ultimate source of values and action" (Schein 2004, 26). Therefore, one conclusion we can draw is that, to enable the changes in the four conditions that in turn will support awakened sustainability, we need to focus more on the transformative change level—that is, the mental models in which systems are grounded. Ultimately, mental models, to use Dan Hill's metaphor (which was presented in Chapter 7 in the context of the four conditions) could be considered as part of the invisible yet impactful "dark matter," which "is the material that absorbs or rejects wider change" (Hill 2012, 84).

Another framework that aligns with the multidimensional nature of systems change is O'Brien and Sygna's (2013) three spheres of transformation model. Their model offers a comprehensive understanding of a transformation "as a process that takes place across three embedded and interacting spheres" (19): practical, political, and personal. The practical sphere focuses on behavior change and technical responses (e.g., technological innovation). The political sphere, which shapes "the rules of the game," consists of "social and cultural norms, institutions, and governance systems that shape behaviors, actions, and investments" (Leichenko and O'Brien 2019, 181). Finally, the personal sphere is "where the transformation of individual and collective beliefs, values and worldviews occur" (O'Brien and Sygna 2013, 21). As in Kania et al.'s model, one level (sphere) seems to be more impactful than the other two—in this case, the personal sphere. As O'Brien and Sygna (2013) note, "Transformations in the personal sphere are considered to have more powerful consequences than in other spheres" (21). At the same time, they also emphasize the connections between the spheres, suggesting that while

---

[2] I find Kania et al.'s (2018) multilevel approach to systems change, as well as how it sheds light on the importance of mental models, to be the most valuable parts of their model, which is why I address these elements in greater detail. At the same time, the other five conditions the model includes (policies, practices, resource flows, relations and connections, and power dynamics) are also reflected in the pathway to change outlined in this chapter.

changes can happen within each dimension, it is the interactions across the different spheres that could generate the most effective outcomes in terms of transformation.

Building on these frameworks for system transformation (as well as others, e.g., the iceberg model) it is clear that the necessary changes in the four conditions will need to be supported by a shift in the mental model dominating corporate sustainability, from "shareholder capitalism 2.0" to "sustainability first, NOW."[3] This transition is not going to be easy, as mental models are quite persistent. As the World Bank (2015) notes, mental models "may outlive their usefulness or, indeed, may persist when they were never useful to begin with. We have a hard time abandoning mental models that are not serving us well" (68). This certainly seems to be the case with the mental models described in the first part of the book (i.e., shareholder capitalism 1.0 and shareholder capitalism 2.0).

The difficulty of changing mental models is directly connected to what Kegan and Lahey (2001) describe as the hidden immune system we have, which resists change. This metaphor is consistent with the notion that transformation processes "often involve struggles and are typically met with resistance related to power, politics, and interests" (Leichenko and O'Brien 2019, 178). Thus, the challenge of changing mental models is intertwined with the challenge of overcoming the system's resistance to change. To do the former, we need to focus on several elements that could be synthesized into a "theory of change" that would result eventually in the desired changes in the mental model as well as in the four conditions (See Fig. 8.2). This theory of change is built on the power of narratives of transformative change to win the "battle of narratives," thus weakening the forces resistant to change, which in return will help build a political alignment of those supporting change, establishing the critical mass necessary to make change happen. This process will eventually lead to the desired changes in the mental model and in the four conditions. Below, I describe each part of this "theory of change" in further detail.

---

[3] See Chapter 6 for further detail.

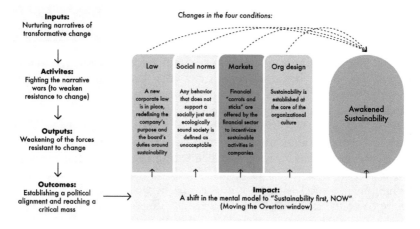

Fig. 8.2   How to shift the mental model dominating business to "sustainability first, NOW" to enable the desired changes in the four conditions

A Theory of Change: Step by Step

1. Inputs: Nurturing narratives of transformative change

One way to change mental models, according to Kania et al. (2018), is to use the power of narratives.[4] Narratives are powerful tools for change-making; Mayer sees narratives as "the most important human device for collective action," and according to Hansen (2020) they create "a bridge from how things are to how they can be" (5). Veland et al. (2018) suggest that "narratives structure human comprehension, and shape our ability to imagine and achieve transformed futures within the 1.5 degree threshold" (41). These researchers point to the role of narratives in enabling a transformative future by framing it in appealing and desirable ways. This point is aligned with Knuth's (2019) research, which finds narratives to be "a key element in deliberate transformation" (58). Knuth defines narratives as "the stories people use to make sense of themselves, their connection to

---

[4] FrameWorks Institute CEO Nat Kendall-Taylor points to the bidirectional relationships between narratives and mental models, whereby "narratives are shaped by mental models, but narratives also, over time, shape the mental models we have" (Kania et al. 2018, 9).

broader group, and to the world as a whole" (58), suggesting that narratives help people make sense of complex systems and understand what should be done to change them.

So, what are the narratives that will ignite the desired shift in mental models?

First, I should clarify that what we need is not one unified narrative, but numerous narratives that, while they share core principles, are constructed in alignment with their specific context (e.g., local versus national) and audience (e.g., company employees versus community members). Furthermore, having in mind the need to create a political alignment of different transformative forces (see "outcomes") suggests the need to develop multiple stories reflecting diverse perspectives. The core principles that will serve as the connecting thread between these narratives are as follows:

1. Meaningful and rewilding—first and foremost, narratives should be meaningful for their audience. Reinsborough and Canning (2017) explain: "Narrative power manifests as a fight over how to make meaning. We often believe in a story not because it is factually true but because it connects with our values or is relevant to our experiences in a way that is compelling" (25). Moreover, narratives should open the doors for new possibilities, rewilding people's imaginations and helping them "to break down the old ways of thinking and to move beyond our current conception of what is and what is not possible" (Knights 2019, 13).
2. Empowering—narratives should provide people with agency and vitality to "stay with the trouble" (Haraway 2016), integrating urgency to act on the climate crisis with the belief that it is not too late to do something about it. They should echo Heglar's (2020) point that "because the thing about warming—whether we're talking about the globe or a fever—is that it happens in degrees. That means that every slice of a degree matters. And right now that means everything we do matters" (281).
3. Well-being centered—fighting climate crisis and social injustice should be framed in terms of personal well-being, creating a clear connection between tackling big issues and improving people's living conditions. The notion of well-being should be broad; as

suggested by the OECD well-being framework, it should incorporate within it "non-material dimensions, including social connections, health, quality of jobs, quality of the environment, etc." (OECD 2018, 5).

4. Impactful in the present, not just the future—benefitting people's well-being solely in the future is not a compelling story for people who struggle with daily hardships now, especially if these future gains represent a trade-off against present concessions. Narratives should therefore focus on positive impacts in the present as much as in the future (e.g., discussing climate change in terms of "good jobs," or immediate pollution reduction in underserved communities).

2. Activities: Fighting the narrative wars (to weaken resistance to change)

The battles over meaning construction happen all the time, whether we are aware of them or not. The shift to a mental model of "sustainability first, NOW" cannot avoid these battles either. First, "narrative fighters," as I refer to those actively participating in these battles, should recognize that they do not operate in a void: "to enact a new vision, we must break free from the control of the old way of doing things" (Hansen 2020, 5). Second, they should have a clear plan for how to win the battle, taking into account strategies that were developed to do so (e.g., Sachs 2012; Reinsborough and Canning 2017; Ganz 2010).

Reinsborough and Canning (2017) argue that "to win the battle of the story requires a narrative designed to connect with your audiences' values, challenge underlying assumptions, and outcompete opposing narratives" (59). They focus on five story elements that are critical to winning this battle: "conflict, characters, imagery, foreshadowing, and assumptions" (59). Sachs (2012), who builds on Joseph Campbell's hero journey, suggests that effective storytellers "clearly define their heroes, villains, and the conflict between them to show how their epic plays out in the lives of characters we can relate to" (25). Finally, Ganz (2010) offers the concept of public narrative, which brings together three stories: a story of self, a story of us, and a story of now. He suggests that public narrative can be used to "move to action by mobilizing sources of motivation, constructing new shared individual and collective identities, and finding the courage to act" (19).

In addition to choosing a strategy, it is also important to consider who the messenger is. As Mayer (2014) notes: "Receptivity to narrative also depends on prior attitudes toward the storyteller" (115). This notion is consistent with the impact of social identity on people's level of openness to messages, as could be seen for example with the influence of Greta Thunberg (i.e., "the Greta Thunberg effect"), which was found to be stronger with liberals than with conservatives (Sabherwal et al. 2021).

3. Outputs: Weakening of the forces resistant to change

The narrative wars should produce initial gains (i.e., outputs) in the form of making the forces opposing (transformational) change weaker. This is an important (and sometimes overlooked, Hansen 2020) aspect of the change process, given the power of forces benefiting from the status quo to prevent or slow change. The emphasis on resistance to change echoes Kurt Lewin's (1951) force field analysis, which considers an existing behavior of a person or organization as a dynamic equilibrium between forces operating in opposite directions (i.e., driving and restraining forces). A shift to a new state of equilibrium can be achieved by increasing the forces driving change and/or diminishing the forces resisting change. Lewin emphasized the need to focus on the latter, as "increasing driving forces often tends to be offset by increased resistance" (Lunenburg 2010, 6).

One example of the need to weaken restraining forces is the long fight of fossil fuel companies against climate action, which has delayed action for decades.[5] Harvard Professor Naomi Oreskes (2019) describes their successful efforts to spread disinformation and block meaningful action on climate change as "the climate-change scam that beat science, big time" (para. 1). Mann (2021) describes the forces opposing action on climate change as "the forces of denial and delay—the fossil fuel companies, right-wing plutocrats, and oil-funded governments that continue to profit from our dependence on fossil fuels" (3). Furthermore, Alex Steffen suggests that these forces apply tactics of "predatory delay"—that is, "the blocking

---

[5] Season 1 of the podcast "Drilled" provides a detailed expose of these efforts. See https://www.criticalfrequency.org/drilled.

or slowing of needed change, in order to make money off unsustainable, unjust systems in the meantime" (McKibben 2018, para. 24). Therefore, these forces should be in a weaker position, with less influence, following the battle of narratives.

4. Outcomes: Establishing a political alignment and reaching a critical mass

Weakening the forces that oppose change will allow for the creation of what Smucker (2017) describes as a "political alignment." According to Smucker, big changes require the formation of a broad political alignment, defined as "an aligning of different social groups, organizations, and movements into a unified force that intervenes to shape social, economic, and political reality" (Smucker 2017, 244). The Sunrise Movement has incorporated this idea into its theory of change, suggesting there is a need to build a New Green Deal alignment that will follow the success of past alignments in the New Deal and Reagan eras (Girgenti and Shahid 2020).

Building on Smucker's model and the Sunrise movement's example, I see a clear imperative for establishing a broad political alignment of actors supporting the vision of a socially just and ecologically sound society. This alignment should include not only political forces (i.e., political parties and movements), but also other elements in society that support a transformative change to fight the climate crisis and social injustice, including companies, communities, social groups, and organizations, among others. To succeed, this alignment must develop a big tent mentality, avoiding purity and encouraging collaboration between different actors, even when they do not see eye-to-eye about anything apart from the need for transformation.

Political alignment is a powerful platform for change, but creating the momentum necessary for shifting the dominant mental model to "sustainability first, NOW" will require not just a platform, but also enough people working to make it happen. This notion reflects the theory of critical mass, which "argues that when a committed minority reaches a critical group size—commonly referred to as a 'critical mass'—the social system crosses a tipping point. Once the tipping point is reached, the actions of a minority group trigger a cascade of behavior change that rapidly increases the acceptance of a minority view" (Centola et al. 2018, 1116). A key question then is what size of critical mass is needed to reach a tipping point (or more accurately a social tipping point, see Milkoreit et al. 2018) in this case?

Studies so far suggest a range of 10–40% of the population as the size of an effective critical mass (Centola et al. 2018). The only study looking into this question empirically found that 25% of the population is the required size in the case of changing social norms (Centola et al. 2018). Chenoweth and Belgioioso (2019) go even further, suggesting that, in the case of mass uprisings against oppressing regimes, it was sufficient to mobilize only 3.5% of the population to achieve a regime change. This suggestion is based on prior research by Chenoweth and Stephan (2011) into nonviolent resistance, later defined by Chenoweth as the "3.5% rule"—"no government has withstood a challenge of 3.5% of their population mobilized against it during a peak event" (Chenoweth 2020, 1).

Climate activists seem to adopt the "3.5% rule" as part of their theory of change (see, e.g., Robson 2019; Prakash 2020), although this approach has been criticized, given some of the limitations of the rule that may make it less applicable to climate activism (Ahmed 2019).[6] While reaching a critical mass is mainly a numbers game, it should be noted that not all activities are necessarily equal in terms of the impact they create. Extinction Rebellion Cofounder, Roger Hallam, suggests for example that impact equals the number of people multiplied by the degree of disruption. "If you have millions of people but they don't create any disruption, you're going to have minimal outcome. You need lots of people causing a reasonable amount of disruption," he observes (Samuel 2020, para. 18). While Hallam's approach may have some merits in the case of Extinction Rebellion, it may be the case that different contexts will require different tactics.

5. Impact: A shift in the mental model, or Moving the Overton window

A key impact we can expect to see when the political alignment is created and the critical mass is reached is a shift in the Overton window, in which

---

[6] Chenoweth herself suggested that the rule derived from—and thus should be applied only to—specific types of campaigns that focus on "overthrowing a government or achieving territorial independence" (Chenoweth 2020, 3).

the mental model of "sustainability first, NOW" moves from being an unthinkable proposition to a sensible one. The metaphor of the Overton window is useful here, as it frames a range of ideas that are considered acceptable on any given issue—these are what is inside the window. Everything else (i.e., what is outside the window) is considered somewhere between radical and unthinkable. According to Bregman (2020), "if you want to change the world, you need to shift the window. How? By pushing on the edges. By being unreasonable, insufferable, and unrealistic" (para. 62). To some extent, this is exactly the strategy to be pursued here.

Climate activists have shifted the Overton window over the last couple of years. Wallace-Wells (2020) gives the examples of "Prakash and AOC, Justice Democrats and New Consensus and Roosevelt Forward and Data for Progress, XR and Greta and Xiye Bastida and Jamie Margolin and Alexandria Villaseñor and many, many more" (para. 32), whose alarmism shifted the Overton window toward political climate action. Similar examples of attempts to move the Overton window can be found in business, where employee activism has been pushing CEOs and boardrooms to take climate change and social issues more seriously and to adopt policies and practices that were, until recently, considered radical or not viable. Examples include employees in tech companies, such as Microsoft, Google, and Amazon, demanding their companies adopt bolder carbon reduction goals, stop selling products to fossil fuel companies, and stop funding politicians and lobbyists denying climate change. Further examples include employees urging their companies to adopt more progressive policies on social issues, including racism, inequality, collaboration with organizations that violate human rights, and so on.

This surge of climate and social activism paves the way for the next waves of "Overton window movers," which will be driven by the developments detailed in the prior steps (outputs and outcomes), further pushing "sustainability first, NOW" from the margins to the mainstream.

Connecting the Dots: Transformation Pathways and Awakened Sustainability

Awakened sustainability could be considered as a response to the question "how might we design business to allow for a socially just and ecologically sound society?" echoing Sharma's (2017) point that "a radical transformer asks questions that require new ways of thinking and acting" (xix). While certainly not the only response, awakened sustainability can be considered what Sharma calls "a conscious full-spectrum response" (5) to the symbiotic relationships that companies have with the key challenges we face. The roadmap presented in Chapters 7–8 provides guidance on how to reach the "promised land" of awakened sustainability, which is described in detail in Chapter 6. Yet, there are two questions that may remain unanswered: Can all the transformation pathways shown in Chapter 5 take us there? And how much time will it take?

First, all the transformation pathways could lead to awakened sustainability (see Fig. 8.3), even "fixing the system," which seems to represent a "lighter" version of transformation in comparison with the two other pathways.[7] While this pathway's current trajectory may seem far from reaching awakened sustainability in the foreseen future, its ambitions could be ratcheted up to bring them increasingly closer to the ones embedded in awakened sustainability. This scenario is built on the emergent nature of changes in complex systems, especially when they interact with possible shifts in the Overton window that could "radicalize" the "fixing the system" pathway. Capitalizing on Wahl's (2016) insights on emergence and design, aligning the trajectory of this pathway with awakened sustainability would also depend on the ability of actors advancing this pathway "to work with unpredictability and emergence rather than against it" (224).

Second, the timeframe to reach awakened sustainability depends on which pathway might become the most dominant along the way. One comparison that may be helpful is for achieving zero emissions, which could be done in different timeframes (i.e., 2040, 2050, 2075, etc.), depending on how aggressive the decarbonization plan and its execution are. The more aggressive the plan and its execution, the less time it will take to reach zero emissions. Similarly, in this case, if "fixing the system" becomes the dominant pathway, then it may take longer to reach awakened sustainability; if one of the other two trajectories takes the lead, the timeframe will be shorter (see Fig. 8.3).

One framework that could be useful in bringing these pieces together is Sharpe's (2020) Three Horizons framework, which presents the dynamics of change across time through three horizons. It shows "the shift from the established patterns of the first horizon to the emergence of new patterns

---

[7] Out of the three pathways "fixing the system" is the one closest to sustainability-as-usual in terms of its approach to changes in corporate sustainability.

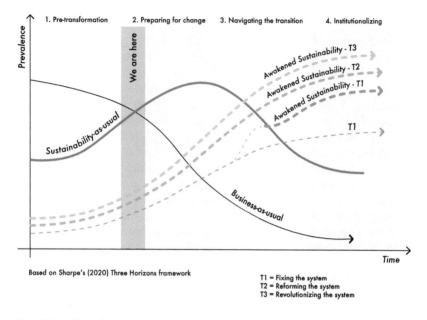

Fig. 8.3   Transformation pathways to awakened sustainability

in the third, via the transition activity of the second" (10–11). Building on this framework, Fig. 8.3 displays the transformation from business-as-usual through sustainability-as-usual to awakened sustainability, along with Moore et al.'s (2014) four phases of transformation processes.[8] The figure brings to view how sustainability-as-usual could evolve into one of the three potential pathways of transformation: fixing, reforming, or revolutionizing the system. We can see in the graph how "fixing the system" has two potential trajectories—a "ratcheted up" trajectory (Awakened sustainability—T1) that leads to awakened sustainability and another one that leads only to a moderate improvement to sustainability-as-usual (T1). The pathways themselves are likely to interact with and affect one another, which could potentially reduce some of the differences between them and make them more alike.

---

[8] See Chapter 5 for further detail on Moore et al.'s (2014) four-phase framework for transformation processes.

Companies will have to consider all of this as they decide on their approach and how and when to respond to the unfolding transformation processes. They may want to consider "the inverse relationships between degrees of freedom and signal strength" (McGrath 2019, 49),[9] which suggests that "...the moment at which you have the richest, most trustworthy information is often the moment at which you have the least power to change the story told by that information" (51). Consequently, companies will need to make a strategic choice based on evaluating which of the transformation pathways gather more momentum. As they do so, they may look for what McGrath (2019) describes as the "period of optimum warning," where there is some balance between the risk and return of responding to the "not yet" (Bloch 1995) state of awakened sustainability.

While this book draws on the principle that it is more effective to change the environment than to change the person (Fuller 1969)—or in this case the company—it should also be noted that, when it comes to the transformation process, companies are not bystanders who just need to wait and see which pathway becomes more dominant. Companies could have direct and indirect impacts, both positive and negative, on the ongoing shift from sustainability-as-usual to awakened sustainability. They can, for example, become warriors in the battle of narratives, support norm entrepreneurs, join other forces that work to advance a social justice agenda, advocate for strong laws on climate action, oppose media outlets and politicians spreading disinformation on climate change, and so on. Additionally, companies can further focus on their organizational design (one of the four conditions needed to realize awakened sustainability), ensuring they design an internal environment that is consistent with the changes in the external one. Furthermore, companies could work on developing products, services, strategies, and business models that are in alignment with awakened sustainability principles to help demonstrate the possibilities offered by this state.

Finally, the shift to awakened sustainability as the context in which companies operate may have become somewhat more realistic due to the COVID-19 pandemic and the subsequent economic downturn. As Milton Friedman (1962) wrote, "only a crisis—actual or perceived— produces real change. When that crisis occurs, the actions that are taken

---

[9] According to McGrath, this model was created by the Futures Strategy Group.

depend on the ideas that are lying around. That, I believe, is our basic function: to develop alternatives to existing policies, to keep them alive and available until the politically impossible becomes the politically inevitable" (xiv). While it may be ironic to quote from Friedman to conclude a book that started with Friedman's contribution to the state we are in today, I find his words echo accurately where we currently stand. From now on, it is up to us to make the desired future of awakened sustainability a reality.

References

Ahmed, Nafeez. 2019. "The Flawed Social Science Behind Extinction Rebellion's Change Strategy." *Insurge Intelligence*, October 28. https://rb.gy/8fnai0.
Bloch, Ernst. 1995. *The Principle of Hope*. Cambridge, MA: MIT Press.
Bregman, Rutger. 2020. "The Neoliberal Era Is Ending: What Comes Next?" *The Correspondent*, May 14. https://rb.gy/elqctm.
Centola, Damon, Joshua Becker, Devon Brackbill, and Andrea Baronchelli. 2018. "Experimental Evidence for Tipping Points in Social Convention." *Science* 360 (6393): 1116–19. https://doi.org/10.1126/science.aas8827.
Chenoweth, Erica. 2020. "Questions, Answers, and Some Cautionary Updates Regarding the 3.5% Rule." Carr Center Discussion Paper Series, no. 2020–005.
Chenoweth, Erica, and Margherita Belgioioso. 2019. "The Physics of Dissent and the Effects of Movement Momentum." *Nature Human Behaviour* 3 (10): 1088–95. https://doi.org/10.1038/s41562-019-0665-8..
Chenoweth, Erica, and Maria J. Stephan. 2011. *Why Civil Resistance Works: The Strategic Logic of Nonviolent Conflict*. New York, NY: Columbia University Press.
Friedman, Milton. 1962. *Capitalism and Freedom*. Chicago: University of Chicago Press.
Fuller, R. Buckminster. 1969. *Utopia or Oblivion: The Prospects for Humanity*. New York: Bantam Books.
Ganz, Marshall. 2010. "Leading Change: Leadership, Organization, and Social Movements." In *Handbook of Leadership Theory and Practice: A Harvard Business School Centennial Colloquium*, edited by Nitin Nohria and Rakesh Khurana. Boston, MA: Harvard Business School Publishing.
Girgenti, Guido, and Waleed Shahid. 2020. "The Next Era of American Politics." In *Winning the Green New Deal: Why We Must, How We Can*, edited by Varshini Prakash and Guido Girgenti. New York, NY: Simon & Schuster.
Hansen, Hans. 2020. *Narrative Change: How Changing the Story Can Transform Society, Business, and Ourselves*. New York, NY: Columbia University Press.

Haraway, Donna J. 2016. *Staying with the Trouble: Making Kin in the Chtulucene*. Durham, NC: Duke University Press.

Heglar, Mary Annaïse. 2020. "Home Is Always Worth It." In *All We Can Save: Truth, Courage, and Solutions for the Climate Crisis*, edited by Ayana Elizabeth Johnson and Katherine K. Wilkinson. New York, NY: One World.

Hill, Dan. 2012. *Dark Matter and Trojan Horses: A Strategic Design Vocabulary*. Moscow: Strelka Press.

Kania, John, Mark Kramer, and Peter Senge. 2018. "The Water of Systems Change." *FSG*.

Kegan, Robert, and Lisa Laskow Lahey. 2001. *How the Way We Talk Can Change the Way We Work: Seven Languages for Transformation*. San Francisco, CA: Jossey-Bass.

Knights, Sam. 2019. "Introduction: The Story So Far." In *This Is Not a Drill: An Extinction Rebellion Handbook*, edited by Clare Farrell, Alison Green, Sam Knghts, William Skeaping, and Extinction Rebellion, 186. Penguin.

Knuth, Katherine. 2019. "How Collectives Drive Deliberate Transformation to Make Progress Toward Sustainability." University of Minnesota.

Leichenko, Robin, and Karen O'Brien. 2019. *Climate and Society: Transforming the Future*. Cambridge, UK: Polity Press.

Lewin, Kurt. 1951. *Field Theory in Social Sciences*. New York, NY: Harper & Row.

Lunenburg, Fred C. 2010. "Forces for and Resistance to Organizational Change." *National Forum of Educational Administration and Supervision Journal* 27 (4): 1–10.

Mann, Michael E. 2021. *The New Climate War: The Fight to Take Back Our Planet*. New York, NY: PublicAffairs.

Mayer, Frederick W. 2014. *Narrative Politics: Stories and Collective Action*. London, UK: Oxford University Press.

McGrath, Rita Gunther. 2019. *Seeing Around Corners: How to Spot Inflection Points in Business Before They Happen*. New York: Houghton Mifflin Harcourt.

McKibben, Bill. 2018. "How Extreme Weather Is Shrinking the Planet." *The New Yorker*. https://rb.gy/vpnbgd.

Milkoreit, Manjana, Jennifer Hodbod, Jacopo Baggio, Karina Benessaiah, Rafael Calderón-Contreras, Jonathan F. Donges, Jean Denis Mathias, Juan Carlos Rocha, Michael Schoon, and Saskia E. Werners. 2018. "Defining Tipping Points for Social-Ecological Systems Scholarship—An Interdisciplinary Literature Review." *Environmental Research Letters*. Institute of Physics Publishing. https://doi.org/10.1088/1748-9326/aaaa75.

Moore, Michele Lee, Ola Tjornbo, Elin Enfors, Corrie Knapp, Jennifer Hodbod, Jacopo A. Baggio, Albert Norström, Per Olsson, and Duan Biggs. 2014. "Studying the Complexity of Change: Toward an Analytical Framework for

Understanding Deliberate Social-Ecological Transformations." *Ecology and Society* 19 (4). https://doi.org/10.5751/ES-06966-190454.

O'Brien, Karen, and Linda Sygna. 2013. "Responding to Climate Change: The Three Spheres of Transformation." In *Proceedings of Transformation in Changing Climate International Conference*, 16–23. Oslo, Norway: University of Oslo.

OECD. 2018. "Elements for a New Growth Narrative."

Oreskes, Naomi. 2019. "The Greatest Scam in History: How the Energy Companies Took Us All." *Common Dreams*, November 11. https://rb.gy/5n9ipr.

Prakash, Varshini. 2020. "People Power and Political Power." In *Winning the Green New Deal: Why We Must, How We Can*, edited by Varshini Prakash and Guido Girgenti. New York, NY: Simon & Schuster.

Reinsborough, Patrick, and Doyle Canning. 2017. *Re:Imagining Change—How to Use Story-Based Strategy to Win Campaigns, Build Movements, and Change the World*. 2nd ed. Oakland, CA: PM Press.

Robson, David. 2019. "The '3.5% Rule': How a Small Minority Can Change the World." *BBC Future*, May 13. https://rb.gy/cuuesl.

Sabherwal, Anandita, Matthew T. Ballew, Sander van der Linden, Abel Gustafson, Matthew H. Goldberg, Edward W. Maibach, John E. Kotcher, Janet K. Swim, Seth A. Rosenthal, and Anthony Leiserowitz. 2021. "The Greta Thunberg Effect: Familiarity with Greta Thunberg Predicts Intentions to Engage in Climate Activism in the United States." *Journal of Applied Social Psychology*. https://doi.org/10.1111/jasp.12737.

Sachs, Jonah. 2012. *Winning the Story Wars: Why Those Who Tell (and Live) the Best Stories Will Rule the Future*. Boston, MA: Harvard Business Review Press.

Samuel, Sigal. 2020. "Extinction Rebellion's Plan to Save the Climate with Civil Disobedience." *Vox*, January 14. https://rb.gy/7suxa4.

Schein, Edgar H. 2004. *Organizational Culture and Leadership*. 3rd ed. San Francisco: Jossey-Bass.

Senge, Peter M. 1990. *The Fifth Discipline: The Art and Practice of the Learning Organization*. New York: Doubleday.

Sharma, Monica. 2017. *Radical Transformational Leadership: Strategic Action for Change Agents*. Berkeley, CA: North Atlantic Books.

Sharpe, Bill. 2020. *Three Horizons: The Patterning of Hope*. 2nd ed. Axminster, UK: Triarchy Press.

Smucker, Jonathan M. 2017. *Hegemony How-to: A Roadmap for Radicals*. Chico, CA.

Veland, S., M. Scoville-Simonds, I. Gram-Hanssen, A. K. Schorre, A. El Khoury, M. J. Nordbø, A. H. Lynch, G. Hochachka, and M. Bjørkan. 2018. "Narrative Matters for Sustainability: The Transformative Role of Storytelling in

Realizing 1.5°C Futures." *Current Opinion in Environmental Sustainability* 31: 41–47. https://doi.org/10.1016/j.cosust.2017.12.005.
Wahl, Daniel Christian. 2016. *Designing Regenerative Cultures*. Axminster, UK: Triarchy Press.
Wallace-Wells, David. 2020. "What Climate Alarm Has Already Achieved." *New York Magazine*, 2020.
World Bank. 2015. "Thinking with Mental Models." In *World Development Report 2015: Mind, Society, and Behavior*. Washington, DC: World Bank. https://doi.org/10.1596/978-1-4648-0342-0.
Wright, Erik Olin. 2010. *Envisioning Real Utopias*. London, UK: Verso.

INDEX

153